SpringerBriefs in Water Science and Technology

More information about this series at http://www.springer.com/series/11214

Yiping Li · Harold Lyonel Feukam Nzudie · Xu Zhao · Hua Wang

Addressing the Uneven Distribution of Water Quantity and Quality Endowment

Physical and Virtual Water Transfer within China

 Springer

Yiping Li
Key Laboratory of Integrated Regulation
and Resource Development on Shallow
Lakes, Ministry of Education
College of Environment
Hohai University
Nanjing, Jiangsu, China

Harold Lyonel Feukam Nzudie
Key Laboratory of Integrated Regulation
and Resource Development on Shallow
Lakes, Ministry of Education
College of Environment
Hohai University
Nanjing, Jiangsu, China

Xu Zhao
Key Laboratory of Integrated Regulation
and Resource Development on Shallow
Lakes, Ministry of Education
College of Environment
Hohai University
Nanjing, Jiangsu, China

Hua Wang
Key Laboratory of Integrated Regulation
and Resource Development on Shallow
Lakes, Ministry of Education
College of Environment
Hohai University
Nanjing, Jiangsu, China

ISSN 2194-7244   ISSN 2194-7252   (electronic)
SpringerBriefs in Water Science and Technology
ISBN 978-981-13-9162-0   ISBN 978-981-13-9163-7   (eBook)
https://doi.org/10.1007/978-981-13-9163-7

This Springer imprint is published by the registered company Springer Nature Singapore Pte Ltd.
The registered company address is: 152 Beach Road, #21-01/04 Gateway East, Singapore 189721, Singapore

Contents

List of Figures

List of Tables

Chapter 1
Introduction

1.1 Introduction

Do you know that about 71% of Earth consists of water body? For all the water, only 3.5% is drinkable, and the rest is salty water (Willians 2014). Unfortunately, this available water is unequally distributed around the world. Some areas such as the Middle East region, northern and central India, northern of China, Bangladesh, Myanmar etc. are in water scarce condition whereas the others such as Brazil, Canada etc. are in water abundance (WaterAid Australia 2018). It is estimated to 40% of World's people who are affected by water scarcity (The Daily Star 2018). *"No matter, how much rich you are, you can't live without water"* (Josh 2007). *Water scarcity* is a global concern. Every continent is affected.

Moreover, the growing economic development observed across the world has induced the need from nations for more natural resources and the generation of pollutants through wastes. It is estimated to 80% of wastewater which is dumped into the environment (WWAP 2017). Globally, this nations' development has led to (a) water quality issue, (b) stress on water resources, and (c) the necessity to find other ways to use and to afford water. Hence, nations have used different approaches to tackle the concern of water scarcity and water quality, e.g. *physical water transfer* and *virtual water flow*.

Physical water transfer is an approach consisting of conveying water through channels from abundant water area to water scarce to fill the water gap between the two areas. Virtual water flow also called *virtual water trade* in some literature is defined as the water embedded in commodities and services moving between exchanging areas (Chapagain and Hoekstra 2008). These approaches are focused on the externalisation of water supply.

This book introduces physical and virtual water transfer as approaches to alleviate water distribution issue and on the other hand improve water quality. The previous two approaches have been thoroughly investigated by using several methods such as

© The Author(s), under exclusive license to Springer Nature Singapore Pte Ltd. 2020
Y. Li et al., *Addressing the Uneven Distribution of Water Quantity
and Quality Endowment*, SpringerBriefs in Water Science and Technology,
https://doi.org/10.1007/978-981-13-9163-7_1

Environmental Fluid Dynamics Code (EFDC), Xinanjiang and mathematical equations derived from Leontief model. These first two models based on simulations are of great importance in so far as each of them integrates modules which enable modelling water flow, sediments transport and water age in rivers, lake, and estuaries. Leontief model gives a detailed quantification of the direct and indirect virtual water flow embedded in commodities. In this current book, water endowment systems pertaining to country, megacity, and lakes are used as examples to illustrate how effective these models are in giving the impacts of physical and virtual water transfer between areas. Moreover, these impacts presented in this book are not only provided by clear sentences, more importantly, they are displayed in figures, histograms and tables. Thus, both physical and virtual water transfer can be easily managed for efficient water use.

References

Chapagain AK, Hoekstra AY (2008) The global component of freshwater demand and supply: an assessment of virtual water flows between nations as a result of trade in agricultural and industrial products. Water Int 33:19–32

Josh (2007) Best slogans for save water awareness and scarcity. https://medium.com/@drdrumaff/best-slogans-for-save-water-awareness-and-scarcity-d6e9b76c5dbd. Accessed 16 November 2017

The Daily Star (2018) 40 percent of world's people affected by water scarcity. https://www.thedailystar.net/world/40-percent-worlds-people-affected-water-scarcity-1549057. Accessed 16 March 2018

Wateraid A (2018) List of water-scarce regions. https://www.wateraid.org/au/articles/wateraid-releases-list-of-water-scarce-regions. Accessed 3 July 2018

Williams M (2014) What percent of Earth is water. Universe Today, 2014–2016

WWAP (2017) wastewater: the untapped resource; facts and figures. UNESCO. http://www.unesco.org/new/en/natural-sciences/environment/water/wwap/facts-and-figures/all-facts-wwdr3/fact-15-water-pollution/

Chapter 2
Comparison of Physical and Virtual Water Transfer

2.1 Introduction

It is well known the various uses of water in our society. Water has played a significant role during the development of human society (Liu et al. 2017). About 71% of the earth's surface is water-covered (The USGS Water Science School 2016). However, water is unevenly distributed. During the last three decades, the Chinese economy has experienced the fastest growth among major nations, and it is now ranked second in the world (Liu et al. 2013). However, China's development is increasingly constrained by limited water resources. As the biggest developing country with the largest population, China has been facing serious water scarcity. There are many studies related to water scarcity in China especially for northern China (Zeng et al. 2013; Crow-Miller and Webber 2017; Liu et al. 2017). To tackle water scarcity within a given area, two major approaches have been used worldwide (Chen et al. 2017). First, transfer physical water from water abundant regions to water scarce regions through water project, for example, the "south-to-north water transfer project" (SNWTP). Secondly, manage water in its virtual form, i.e. embodied in the production of goods and services.

Physical water transfer projects are usually executed by government at a regional/country level. The projects are simply determined by transferring physical bulk water from one water abundant area to another area (generally water scarce) where water is in need (Islar and Boda 2014). The "virtual water" concept was first introduced by Allan (Allan 1994) when studying the option of importing food as opposed to physical water to partly solve the water scarcity problems in the Middle East. The virtual water is defined as the volume of water required to produce commodities or services (Liu et al. 2009). Considering this concept of virtual water, there is a water savings option for a given area which imports water intensive products instead of producing them (Zhao et al. 2015a). These two solutions aiming to tackle water scarcity have attracted much interests (Peng et al. 2014; Sun et al. 2016; El-Sadek 2009; Zhang et al. 2017a). However, most studies have focused solely on either physical water transfer or virtual water flow, few studies have combined and compared the two solutions. Moreover, studies on both physical and virtual water transfer have mainly focused on water quantity but ignoring water quality issues (Zhao et al. 2015b). In the following parts, physical and virtual water transfers are presented and then compared.

2.2 Review on Physical Water Transfer Projects and Their Impacts

2.2.1 Water Transfer Projects in the World

Physical water transfer project has a long history since humans began to respond to water scarcity and irrigation demand. It dated back to the era of Egypt's civilization, as ancient Egyptians intended to satisfy demand for irrigation and shipping. In today's area of south Ethiopia, King Menes commanded to build what we know to be the first water transfer project in the world (Zhuang 2016). It transferred water in the Nile to irrigate the land along the channel and eventually facilitated the development and prosperity of Egypt's civilization. It is in the nineteenth century that numerous advanced water transfer projects emerged. Table 2.1 provides a list of particular (the list is non-exhaustive) water transfer projects in the world. Numerous water transfer projects have been built around the world. They all present a common feature, i.e. transferring water from water abundant area to water scarce area. On the other hand, most of them were built for irrigation and electricity generation purposes. Rare is a water transfer project built mainly for ecological protection (for example that of Germany as shown in Table 2.1)

2.2.2 The South-to-North Water Transfer Project in China

China feeds approximately 21% of the world's population with only 6% of the world's total water resources and 9% of the world's arable land (Liu et al. 2013). This achievement has be performed through considerable investments on water conservancy projects, especially water transfer projects, dams, reservoirs and irrigation infrastructures.

In the beginning of 1950s, Chairman Mao Zedong proposed to perform a water transfer project, the South-to-North water transfer project (SNWTP). This idea arose after he inspected the Yellow River basin in 1953 and suggested "borrow some water from the Yangtze River to the Yellow River" (Liu and Zheng 2002). In the late

Table 2.1 Summary of water transfer projects around the world (Feukam et al. in processing)

Water projects	Characteristics	Timeline	Water transferred	Purposes	References
Snowy River Scheme in Australia	Less than 100 km 16 large dams, 7 hydropower stations, over 145 km of tunnels and about 80 km of aqueducts	Started in 1949	1.1×10^3 m^3/ yr.	Hydropower and Irrigation	(Pittock et al. 2009)
Agus—Segura Transfer in Spain	5 dams, 286 km of main pipe	1978 completed	0.6×10^3 m^3/ yr.	Irrigation and urban water supply	(Pittock et al. 2009)
The Great Man-Made River Project in Libya	Length 3500 km	1980s	2×10^9 m^3/ yr.	water supply for households, agriculture, and industry	(Qadir et al. 2007)
Highlands Water Project between Lesotho and South Africa	200 km of tunnel	Conceived in 1950s and formalized in 1986. Planned as four phases, Phase I was completed in 2003, and Phase II was approved for construction in 2008	0.6×10^3 m^3/ yr. (Phase I only)	Water supply for South Africa's Gauteng industry region. Electricity, royalties and infrastructure for Lesotho.	(Pittock et al. 2009)
water transfer projects in California (USA)	Length 900 km	–	5.2 billion m^3/yr.	multipurpose development project in America to solve the condition of waterlogging in the northern region and drought in the south	(Zhan et al. 2015; Zhuang 2016)

(continued)

Table 2.1 (continued)

Water projects	Characteristics	Timeline	Water transferred	Purposes	References
Rio Sao Francisco Project in Brazil	Total length is about 720 km. Northern axis: 4 pumping stations,22 canals, 6 tunnels, 26 small reservoirs 2 hydroelectric plants of 40 megawatts and 12-megawatt capacities Eastern axis: 5 pumping stations; 2 tunnels and 9 reservoirs.	started during the colonial period and was taken up again by President Lula de Silva in 2000	26–127 m^3/s	Public supply and multiple uses, mainly for irrigation of about 330 000 ha. Bring 2 092 km of dry riverbeds back to life.	(Pittock et al. 2009)
A huge river-interlinking project in India	The project consists of 30 links and some 3000 storage points, connecting 37 rivers for a length of about 4440 km	–	178 × 109 m^3/yr.	generating electricity, increasing irrigated areas and creating a significant fishery	(Qadir et al. 2007)
South-to-North Water Transfer Project in China	About 3833 km Planned as three phases.	Approved in 2002 by the Chinese government	44.8 billion m^3/yr.	supplementing water sources to the North and northwest regions of China	(Zhao et al. 2015a)
Bavaria Water Transfer Project in Germany	–	–	0.15–0.3 billion m^3/yr.	Ecological protection.	(Zhuang 2016)

1950s, both the China Academic Scholars (CAS) and the Ministry of Water Resources began to execute integrated survey work in the upper reaches of the Yangtze River and the southwestern rivers for the possibility and feasibility of transferring water to upper Yellow River areas (Liu and Zheng 2002). In 1972, approximately seven provinces from northern China faced a huge drought (Liu and Zheng 2002). However, it is in the 1990s that frequent droughts occurred and led to severe water shortages in Huang–Huai—Hai River Basins. To combat the drought, Premier Zhou Enlai suggested elaborating a plan to divert lower Yangtze River water to the North China Plain. In 2002, the Chinese government approved the ambitious SNWTP that would take approximately 40 years to construct and would cost more than 58 billion US dollars (Cheng and Song 2009). The project is expected to divert water from the Yangtze River to the North through three routes, and may improve the degraded ecosystems of approximately 151000 km^2 area. The construction of the east and middle routes started in 2002 and 2003 respectively. The west route is in the stage of re-evaluation (Cheng and Song 2009). Globally, the SNWTP (its magnitude is even greater than the Three Gorges Dam Project) would deliver about 45 billion m^3 of water from the Yangtze River to the water starving North China Plain (He et al. 2010).

2.2.3   Water Transfer Projects for Water Quality Improvement

The diverted water derived from water transfer projects is used to meet social needs namely for drinking, supplying agricultural production and industry. On the other hand, it is also used for environmental remediation by flushing out (diluting) pollutants from polluted water bodies (Ghassemi and White 2006; Manshadi et al. 2015). With the growing water needs from the population, Water quality issue has drawn much attention. In 1990s, the Chinese government took several measures to control the pollutant discharge and to enhance the water quality of Taihu Lake, which is the third largest lake in China (Yang and Liu 2010). Eight years later, Taihu Lake was classified by China's central government as a top environmental priority. About US $10 billion was invested by the Chinese government in the pollution control, flood control projects, and pumping stations for the lake. The pumping stations located in specific points of the Taihu Lake's rivers network assure water balance into the lake during the dry season by pumping water in/out of the lake to Yangtze River. In a similar line, water transfer project consisting of two routes and including pump stations was carried out to transfer water from the Yangtze River to Chao Lake to restore the water quality of the lake. This project transfers about 1 billion m^3 of water annually into Chao Lake.

Studies have focused on the effects of physical water transfer on water dilution. Considering water bodies in China, Li et al. (2011) used a three-dimensional numerical model called Environmental Fluid Dynamic Code (EFDC) to show the impacts of water transfer from the Yangtze River to Lake Taihu. They found that the pattern of water age is highly influenced by inflow/outflow tributaries and wind including its magnitude and direction. Li et al. (2013) have developed the concept of water age and Lagrangian particle tracking to assess the appropriate transferred inflow rates, and the environmental impacts of the different water transfer routes on both Lake Taihu (receiver) and the Yangtze River (supplier). They found that the Yangtze River diversion, as an emergency stop-gap measure played important roles in enhancing water exchange in the lake and had minimal impact on the Yangtze River. Huang et al. (2015) used the hydrodynamic-phytoplankton model to simulate the effects of water transfer on Taihu Lake. They found that water transfer has a high impact on it, even so, some factors such as wind conditions, amount and quality of water inflow, and chlorophyll-a distribution are more relevant variables in alleviating algal blooms. Gu et al. (2017) used an integrated hydrodynamic and water quality model to argue that during water transfer, diverted water flow, diverting time, diverting points, sluice condition and sewage interception played essential roles in Chao lake's water quality improvement. Huang et al. (2016) Coupled EFDC and Xinanjiang models to simulate the impacts of water transfer from the Yangtze River on Lake Chao. Using the concept of water age and resident time, they found that the water transfer has enhanced water dilution capability and pollutants removal of the lake Chao. They also found that wind conditions and water transfer routes acting together properly would significantly affect the improvement of the water transfer benefits.

2.2.4 Multidimensional Impacts from Water Transfer Project

Physical water projects have multi-level impacts. The project would present many benefits such as; (a) increasing development rate (with the increasing amount of water, the region can increase its crops production, and then develops its industries); (b) employment opportunities (the project would generate many employments such as monitoring, maintenance); (c) flood control (using the dimensions of the channel and the presence of pumps, the amount of transferred water can be adequately estimated and controlled); (d) navigation (the water transfer channel can be used like a waterway to carry products from one point to another and vice versa); (d) irrigation (groundwater has been used as the main source in agriculture for a long time, but with the surplus from water abundant area, irrigation can be improved); and (e) electricity generation through dams, recreation, tourism, and aquaculture. Water transfer projects may also have a beneficial impact on relieving desertification to a large extent and have a greater impact on regional climate (Magee 2011).

On the other hand, however, water transfer projects have also been the causes of significant environmental and socio-economic problems, such as, (a) loss of land and riparian habitat (the huge canals caused by engineering works have destroyed land). This induces the displacement of people, the loss of community identity, and even the loss of a nations' cultural heritage (Islar and Boda 2014) (for their safety, people close to the channel routes should be displaced). (b) Change in hydrology of river systems (natural flow, especially for the receiving river may change). (c) Alteration of scenery and introduction of new aquatic plants and animal species from the donor basin and damage to fisheries and wildlife species. Water is the habitat for many organisms. Hence, transferring water through the canal can induce their migration. Since the receiving area does not have a similar climate as the supplying area, some species may become threatened). And (d) reservoir-induced seismicity and water-borne diseases (He et al. 2010). In addition, the water salinity changes due to seawater transfer intrusion inducing changes of the groundwater system, soil moisture, and water table (Huang and Pang 2010). In particular cases, the reduction of water source (donor basin) may potentially lead to changes in the natural flow regime, diminish its ability to assimilate pollutants and to support habitat for native aquatic communities, and reduce its availability to provide water-based recreational activities and aesthetic qualities (Zhang et al. 2015). Despite its usefulness, that is to say, increase water supply for agricultural, residential, commercial, hydropower, and other demands, the implementation of water transfer projects has always been an emotional and controversial issue. One of the main reasons is water loss during the process (Ma et al. 2016), which mainly includes evaporation, seepage, bank storage and those impacts mentioned above.

2.2.5 Water Transfer Projects and Their Water Availability Related to Climate Change

Aforementioned, water transfer project is a means for mitigating water scarcity in some areas, although some variables such as population growth, climate change, and environmental constraints could impede the sustainability of the water transfer project (Dargan and Carver 2011). Therefore, it matters to understand how water resources are affected to avoid water scarcity in the supply area and, on the other hand, to meet water needs in a relatively long term. That is one reason many researchers have been interested in the relation between water transfer projects and climate change. Shrestha et al. (2015) predicted future temperature and precipitation of Melamchi basin in Nepal to show that despite increasing water availability expected, Kathmandu Valley would still be facing water scarcity in the future. Li et al. (2015) concentrated on the upper Han River and Luan River supplying and receiving area respectively of the middle route of the SNWTP. They showed that under modelled future scenarios, the combined action of climate change and land cover change to 2050 could decrease the mean annual inflows to the Danjiangkou reservoir of the upper Han River catchment by up to $63.8 \times 108\,m^3$. On the other hand, the Luan River flow could also decrease by up to $18.3 \times 108\,m^3$ or increase by up to $9.3 \times 108\,m^3$ controlled by the direction of land cover change. Consequently, they could impact future water resources availability. Ashoori et al. (2015) focused on California State and found that the projection of future climate change (temperature, precipitation) is likely to affect water availability of the principal water suppliers of Los Angeles County namely Los Angeles, California, and Colorado River Aqueduct. Zhang et al. (2017a) focused on the SNWTP, they took precipitation and temperature as climatic variables and showed that any changes of these variables under certain scenarios could significantly affect water availability of the donor area in return threaten the project benefits.

2.3 Review on Virtual Water Transfer

"Virtual water" was termed by Allan in the early 1990s to interpret the amount of water input used to generate agricultural products such as cereals and livestock (Allan 1998). Later, the theory of virtual water was developed, and the terminology and scope of virtual water were extended to indicate the water required by the production of goods and services (Fulton et al. 2014; Mekonnen and Hoekstra 2011). It is also termed as "embodied water", and closely correlated with the concept of water footprint (Chapagain and Hoekstra 2004; Hoekstra and Chapagain 2006; Shrestha et al. 2013). In China, Prof. Cheng Guodong first introduced the concept of virtual water and believed that virtual water theory could promisingly ensure water security in China (Zhang et al. 2010). Virtual water is considered as an important indicator in assessing the dependency of a given region/country on its water resources and in revealing water balance in trade. Since the introduction of the concept, virtual water has been extensively used as a tool to quantify the real water demand of various economies in formulating the water shortage problems in many regions, especially when evaluating water dependency and assessing water policies (Chen and Li 2015; Chapagain et al. 2006; Novo et al. 2009). Virtual water has drawn growing attention among policymakers, scientific communities, and the general public to research on quantifying virtual water content, virtual water trade, virtual water trade patterns, food security, and virtual water management and so on (Sun et al. 2016).

2.3.1 Virtual Water Accounting Approaches

There are two major approaches employed to evaluate virtual water (Feng et al. 2014); the *bottom-up* approach and the *top-down* approach usually based on input-output analysis. The former approach estimates virtual water flows by calculating the virtual water content of goods (water used throughout the production process of a good) and associated international trade from detailed trade data (Feng et al. 2014). It has become one of the most popular approaches in virtual water and water footprint studies due to its relatively good data availability. Hoekstra and Mekonnen (2012) used a bottom-up approach to quantify virtual water flows (with the differentiation of green, blue and grey water) related to international trade in agricultural and industrial products through the concept of water footprint. They estimated the total volume of virtual water trade at 2320 Gm^3/y. Duan and Chen (2016) used a bottom-up method to study virtual water flows embodied in the international energy trade of China and showed that during 2001–2014 China was a net virtual water importer. The bottom-up does not make a distinction between intermediate and final users, in terms of water consumption. Therefore, it cannot comprehensively describe supply chain effects, which are crucial for allocating responsibility to the final consumer and identifying driving forces.

The Environmental input-output analysis as a top-down approach calculates virtual water and water footprint via tracing the whole regional, national, or global supply chain depending on the accounting framework used. In the top-down approach, water consumed in production is allocated to final rather than intermediate consumers. Zhao et al. (2009) used input-output to calculate national water footprint, and found that China was a net virtual water exporter regarding the whole sector in 2002. Lenzen et al. (2013) adopted the international input-output model to study the global flow of virtual water. They found that developed countries imported an increasing amount of virtual water, which to some extent alleviated their water scarcity problems.

Other authors have rather used a top-down approach. Chen et al. (2017) applied the input-output model to quantify water footprint for each province and inter-provincial virtual water in China. They found that virtual water was mainly transferred from Northwestern, Northern, and Northeastern China to developed coastal regions in Eastern and Southern China. This indicates that the water-deficient regions transferred a significant amount of virtual water to the water-affluent regions. Feng et al. (2014) also used a multi-regional input-output approach and the water stress index indicator to evaluate virtual water flow and its associated impacts amongst 30 provinces in China. According to their studies, Xinjiang, Hebei, Inner Mongolia, and Jiangsu are the top virtual water exporting provinces compared to other provinces in particular Shandong, Shanghai, Tianjin, and Zhejiang. They also mapped areas representing ecosystem impacts due to virtual water flow and showed that Xinjiang, Hebei, Inner Mongolia are the most affected. The major problem with the top-down approach lies in the aggregation of processes and products at the level of economic sectors and the relatively high aggregation level especially of different agricultural sectors due to the given data in national accounts.

2.3.2 Virtual Water Trade and Its Impact on Water Scarcity

In addition to quantification of virtual water, various cases have addressed the connection between virtual water trade and regional water scarcity. Arid countries in the Middle East and North Africa, for example, can reduce substantially blue water usage by importing food, which corresponds to importing water (Roson and Sartori 2010; Yang and Zehnder 2002). Yang and Zehnder (2001) concentrated on China and argued that virtual water import should be incorporated into water policy to alleviate water shortage. Wichelns (2001) took Egypt as an example to show the importance of water trade embodied in food commodities in the water-scarce region. Roson and Sartori (2010) and Fracasso et al. (2016) both chose Mediterranean regions to illustrate the virtual water flow among these countries. They pointed out that the virtual water trade in agriculture among these countries achieved a reasonable allocation of water resources and alleviated the water scarcity in relevant countries. Biewald and Rolinski (2012) reacted to some critics of virtual water and argued that, it is a useful concept, and water-scarce regions can alleviate their scarcity through international trade. Studies suggested that some areas are so dependent on importing water that they simply could not sustain the population without it. Jordan, for example, imports annually around 5–7 × 10^3 m^3 of water in virtual form (Porkka 2011). Imports are imperative for compensating water resource deficiency, and the associated embodied virtual water becomes a fundamental alternative source of water for water-scarce countries. Regions with little freshwater resources are capable through trade, to consume more water than that which is locally available (Reimer 2012). At last, research studies focusing on virtual water have largely viewed food trade as a meaningful tool for mitigating the stress caused by water required for food production (Kumar and Singh 2005)

2.3.3 Global Water Savings from Virtual Water Trade

Professor Allan elaborated on the idea of using virtual water import as a tool to decrease the pressure on the scarcely available domestic water resources (Allan 1998). This concept took on more practical meaning once quantification and calculation of virtual water flows have been done. Many types of research introduced the water trading as a water-savings option in many parts of the world such as Southeast England (Yu et al. 2010), China (Zhao et al. 2009) and Australia (Ridoutt et al. 2009). Despite being only one of the many factors of agricultural production and trade (other factors include economy, labour, agricultural land, etc.), water-savings resulting from trade has been a core of interests.

Fraiture et al. (2004) is one of the few studies that show empirically that, when cereals are exported from water abundant to water-stressed countries, such as those in the Middle East region, consequently, local (and global) water 'saving' results. Chapagain et al. (2006) calculated the volume of global water-savings from the international trade of agricultural products and argued that it is equal to 352 Gm^3/yr. (during 1997–2001). Konar and Caylor (2013) evaluated virtual water flows savings in food commodities within Africa, and they found that it is equal to 9.14 × 10^3 m^3. They also showed that the most water savings within Africa occur between South Africa and Mozambique. Dalin and Rodríguez-Iturbe (2014) found that in 2000, global water savings represented 4% of the water used in agriculture. Moreover, trading in agricultural commodities is increasingly being viewed as a useful mechanism for redistributing water in relatively large quantities, and particularly saves water in the country of import (Kumar and Singh 2005).

2.3.4 Virtual Water Trade as a Food Security Tool

Trade in agricultural commodities allows region/country to diversify their diet by sharing their food. Since agriculture is the largest water user, a substantial amount of studies have focused on the agricultural virtual water trade and its relation to improve food security. For example, Novo et al. (2009) and El-Sadek (2011) took Spain and Arab state (especially Egypt) respectively as examples to study the effects of international trade of agricultural products on food security in these areas with virtual water embedded in food as yardstick (Chen et al. 2018). They argued that virtual water has led to improve food security in these regions. Yang and Zehnder (2007) showed that the research on virtual water is linked to food security, thus, strengthen the studies on water-food nexus. Yawson et al. (2013) argued that under some requirements, virtual water is an effective mechanism for enhancing food security. Konar and Caylor (2013) focused on Africa. They used an empirical analysis to show that increasing virtual water trade is correlated with enhancing food security. However, the export of water-intensive products contributes to the loss of significant amounts of domestic water resources.

2.3.5 Virtual Water as an Efficient Water Use and Water Policy Tool

The implementation of a sustainable virtual water trade tends to be considered as one solution to further water use efficiency in the light of existing water scarcity, global, and regional imbalances in water availability. In a similar line, several researchers argued that water-scarce regions can achieve high global water use efficiency by importing products that have high virtual water content embedded in them and exporting products that have very low water content embedded (Kumar and Singh 2005), while the underlying rationale in the virtual water trade argument is "*global water use efficiency*" and "*distribution of scarcity*". There are different schemes to foster water-use efficiency in the agricultural sector (i.e. mechanisation, water-savings, irrigation, and fertilizers, etc.). Dalin and Rodríguez-Iturbe (2014) argued that virtual water trade through agricultural commodities is a way to improve global water-use efficiency by virtually transferring water resources to water-stressed regions.

In the discussion on the implications of virtual water trade as a policy option for countries/regions, it has been pointed out that virtual water trade should be encouraged to promote water savings for arid countries and at a global level through enhancing food security through appropriate and fair trade agreements (Gualtieri 2009). For instance, Biewald (2011) argued that virtual water likely leads to wrong policy implications, however, in some cases when considering not only the consumption side of the virtual water but the production side as well, it can help to develop water policies. Chapagain and Hoekstra (2008) also argued that virtual water could be used as a tool by governments to release the pressure on their water resources.

2.3.6 Driving Forces of Virtual Water Trade

Numerous potential factors influencing virtual water trade are presented in literature. However, in this study, we have highlighted the most relevant drivers. In the first place, water endowment is one driver which can constrain a region to involve in trade of water-intensive products particularly agricultural products. In the second place, there are factors more pertinent than water endowment which should be considered such as socio-economic variables including trade policy and irrigation cost. Chapagain and Hoekstra (2008) also corroborated the fact that international trade (in the form of virtual water) in agricultural commodities depends on much more factors other than water, such as availability of land, labour, knowledge and capital, competitiveness (comparative advantage) in certain types of production, domestic subsidies, export subsidies and import taxes. Among the socio-economic variables, Fracasso et al. (2016) argued that production technologies, domestic and international good prices, trade barriers, national income, irrigation water charges, and other country-specific socio-economic aspects are major drivers of virtual water trade. Tamea et al. (2014) showed that population, gross domestic product, and distance are the fundamental controlling factors of virtual water trade, both for import and export. D'Odorico et al. (2012) explained that geographic proximity among regions is the main driver which defines virtual water trade structure of community. In some cases and according to the nature of the products, quality (good or bad) can be a driver of virtual water flows. It should be noted that water-rich countries export specific goods that, for climatic or commercial reasons, can only be grown there and not in the importing countries despite water abundance, or there are political/commercial agreements that drive goods exchange and are not related to water availability (Tamea et al. 2014).

Note that virtual water trade in commodities occurring between region/country has failed to completely tackle the issue of the uneven distribution of water resources described above, and has even aggravated this adverse situation. This has led to worse water shortages in some water-deficient regions. It should also be noted that the insight into the concept of virtual water is still controversial. Amid the authors who have done researches on it, debate is still rising, Merrett (2003), Ansink (2010), and Wichelns (2015) are of the opinion that virtual water cannot be a driving force to alleviate imbalance water distribution, whereas Sun et al. (2016), and Zhang et al. (2017b) considered it as an effective and helpful measure to mitigate water shortages.

2.4 Comparison Between Physical Water Transfer and Virtual Water Flow

Physical water transfer and virtual water flow may have the same purpose (reduce water scarcity to a large extent) and indeed present similarities, but it should be noted that they have substantial differences. Table 2.2 summarises the differences observed between them. Some points are stressed.

Table 2.2 Differences between physical water transfer and virtual water flow (Feukam et al. in processing)

	Physical water transfer	Virtual water
Scope	Carried out by governments Can be managed at a national level	Concept is termed by Allan Can be evaluated at a provincial, national level or worldwide
Models used		Input-output (multi regional input-output) Bottom-up
Amount	Small	Large
Transferred element	Direct (water)	Indirect (goods, services)
Length	It is geographically limited by position	It does not depend on the geographic position
Benefits	Water Flow can easily be managed	Easily transportable from one place to another place through goods More flexible
Drawbacks	Expensive to execute Difficult to realise due to its intricate engineering works Monitoring is needed for the transferring water channel and pumps to avoid damages Environmental-geological problems such as water quality, slope stability of swelling clay and rock, soil salinisation (Shao et al. 2003)	Need to integrate water policy, technology efficiency Need many parameters to evaluate Difficult to manage (quantify) Indirect impacts on the environment
Sustainability	It is a short term solution, so it cannot be considered as sustainable	More likely to be sustainable

The practice of "virtual water trade" exists since people have engaged in trade relations. In the era of Roman civilization, there was a share of food commodities within Roman area. This was intrinsically followed by a relatively large share of water embeded (Vos and Boelens 2016) as to physical water transfer, it was developed after the trade of goods. Considering the definitions of physical water transfer and virtual water trade given previously, first physical water transfers were implemented *willingly* to respond directly to water shortages, while virtual water trade was used *unconsciously*, that is, people were engaged in trade for meeting their food needs instead of water needs (in doing so, they externalised their water supply).

The cost of physical water transfer is stable and huge about 253 billion Yuan (Chinese currency, 1 Yuan = 0.16 U.S. dollar) (Yang et al. 2015), but virtual water transfer cost is relatively lower. This is due to the fact that trading products also means trading water. The cost of virtual water is flexible due to its component (trade) which sensibly depends on some factors including population behaviour and the dynamic character of virtual water network (Carr et al. 2012). The more population increases, the more water is required to meet needs. However, the amount of Physical water transfer is quantified for some people and for a given time so cannot be flexible with time, whereas virtual water flow is more flexible, because it evolves with the population needs. It is noteworthy that virtual water trade, on the contrary of physical water transfer, may lead a region/country which deals with in economic issues such as price shocks and manipulation of trade policies. Furthermore, in the case of some countries, trade means involvement in a trade network led by powerful interests (Roth and Warner 2007). For example, the Arab States consider that relying on the importation of food commodities is a means to be under foreign domination (El-Sadek 2011).

As aforementioned large amount of funds is required to implement this sort of project (SNWPT for example), so water transfer projects can only be handled at a regional/national scale and by some specific nations which are economically viable. Consequently, it is nearly difficult to be executed by "developing countries". On the contrary, virtual water flow does not need any funds for implementation, because it is naturally into the trade of commodities. Moreover, virtual water more often flows among countries around the world regardless of their water scarcity.

In terms of quantification of the water transferred, physical water transfer and virtual water trade are respectively considered as direct and indirect water transfer. The former is easier to quantify than the latter one. This can be explained by the existence of water pumps in the canal, its dimensions, and the well-known water flow. On the other side, quantifying virtual water transfer is a more complex task. This can be ascribed to its relatively large amount of influencing factors such as source of water to produce goods, data related to trade (import-export), place of production, nature of product (crops, livestock, services) and the like.

Physical water transfer from one point to another requires a lot of structural changes both of the environment and landscape. Physical water transfer project can be implemented or even cancelled according to the presence of natural impediments on its way such as mountains, gulches, etc. This may lead to shorten the length of physical water transfer. The length of physical water transfer is usually shorter than that of virtual water flow. Furthermore, the support for transferring physical water (canal) is fixed because it has been excavated in the ground whereas the support of virtual water (goods or services) flows through trade, so it can easily move far away between the exchanging points (from the source area to the receiver).

After being implemented, physical water transfer needs more attention to avoid some damages (seepage, flooding), for example, due to the depositing which can restrict the opening of the canal, monitoring is required to reduce sludge particularly on the verge. Technical equipment such as water pumps, mechanical gates, and other mechanical equipement presents on-site should be systematically maintained, while virtual water flow is controlled by trade through national or international water policies and sometimes by other factors (described in previous section).

The water which has been transferred during the process of physical water transfer may be of poor quality, and it also brings new species in the receiving ecosystem, thereby, leading to direct impacts on it. Trade in commodities (virtual water) impacts mostly in the supplying area, and its drawbacks are usually more severe in the supplying than in the receiving area.

At last, physical water transfer can be considered as a *"hard path"* to address the issue of water shortages and virtual water a "soft path" (Gleick 2003). The two suggested solutions, however, have different approaches to tackle the issue of water shortages. The "hard path" focuses on finding new water sources and how to transfer it from water abundance to water-scarce areas whereas the *"soft path"* seeks to improve the water use efficiency by providing commodities and services which match users' needs rather than just delivering amount of water as the *"hard path"* does (Gleick 2003).

2.5 Conclusion

The quest for development is still rising. Every nation goes towards a high degree of advancement. This leads to the use of more natural resources. Unfortunately, there is an irregular distribution of water across the world. The over-exploitation of water resources owing to human activities (water-intensive activity) in regions/countries, especially those which have a poor water endowment, leads nations to adopt various ways to use their water. Physical water transfer and virtual water trade are then presented as two solutions to redistribute water. Regions/countries can externalise their water supply from water abundant area. In doing so, physical water transfer and virtual water trade are presented as a water savings tool and allow regions/countries to use water efficiently. However, a close analysis of both physical water transfer and virtual water trade shows that each approach for alleviating water scarcity in one region has indeed played its role (several benefits) to some extent, but has also generated some issues. So both meticulous combine can be an ideal pair for mitigating water stress for a durable development.

References

Allan JA (1994) Overall perspectives on countries and regions. In: Rogers P, Lydon P (eds) Water in the Arab world: perspectives and prognoses. Massachusetts, Cambridge, pp 65–100

Allan JA (1998) Virtual water: a strategic resource global solutions to regional deficits. Ground Water 36(4):545–546

Ansink E (2010) Refuting two claims about virtual water trade. Ecol Econ 69:2027–2032

Ashoori N, Dzombak DA, Small MJ (2015) sustainability review of water-supply options in the los angeles region. J Water Resour Plan Manag 141(12):A4015005

Biewald A (2011) Give virtual water a chance! an attempt to rehabilitate the concept. Gaia 20(3):168

Biewald A, Rolinski S (2012) The theory of virtual water why it can help to understand local water scarcity. GAIA-Ecol Perspect Sci Soc 21:88–90

Carr JA, D'Odorico P, Laio F, Ridolfi L (2012) On the temporal variability of the virtual water network. Geophysical Research Letters 39(6)

Chapagain AK, Hoekstra AY (2004) Water footprints of nations

Chapagain AK, Hoekstra AY (2008) The global component of freshwater demand and supply: an assessment of virtual water flows between nations as a result of trade in agricultural and industrial products. Water Int 33:19–32

Chapagain AK, Hoekstra AY, Savenije HHG (2006) Water saving through international trade of agricultural products. Hydrol Earth Syst Sci 10(3):455–468

Chen GQ, Li JS (2015) Virtual water assessment for Macao, China: highlighting the role of external trade. J Clean Prod 93:308–317

Chen W, Wu S, Lei Y, Li S (2017) China's water footprint by province, and inter-provincial transfer of virtual water. Ecol Indic 74:321–333

Chen W, Wu S, Lei Y, Li S (2018) Virtual water export and import in china's foreign trade: a quantification using input-output tables of China from 2000 to 2012. Resour Conserv Recycl 132:278–290

Cheng S, Song H (2009) Conservation buffer systems for water quality security in south to north water transfer project in China: an approach review. Front For China 4:394–401

Crow-Miller B, Webber M (2017) Of maps and eating bitterness: the politics of scaling in China's south-north water transfer project. Polit Geogr 61:19–30

D'odorico P, Carr J, Laio F, Ridolfi L (2012) Spatial organisation and drivers of the virtual water trade: a community-structure analysis. Environ Res Lett 7:034007

Dalin C, Rodríguez-Iturbe I (2014) Water for food: evolution and projections of water transfers through international agricultural trade. Complexity and Analogy in Science: Theoretical, Methodological and Epistemological Aspects: 248

Dargan SC, Carver SWB (2011) Interbasin transfers of water. In: Hall Booth Smith & Slover, P. C., Atlanta, Georgia 30303 edn. Proceedings of the 2011 Georgia Water Resources Conference, University of Georgia

Duan C, Chen B (2016) Virtual water embodied in international energy trade of China. Energy Procedia 88:94–99

El-Sadek A (2009) Virtual water trade as a solution for water scarcity in Egypt. Water Resour Manag 24:2437–2448

El-Sadek A (2011) Virtual water: an effective mechanism for integrated water resources management. Agric Sci 02(03):248–261

Feng K, Hubacek K, Pfister S, Yu Y, Sun L (2014) Virtual scarce water in China. Environ Sci Technol 48:7704–7713

Fracasso A, Sartori M, Schiavo S (2016) Determinants of virtual water flows in the Mediterranean. Sci Total Environ 543:1054–1062

Fraiture CD, Cai X, Amarasinghe U, Rosegrant M, Molden D (2004) Does International Cereal Trade Save Water? The Impact of Virtual Water Trade on Global Water Use. Comprehensive Assessment of Water Management in Agriculture

Fulton J, Heather C, Peter HG (2014) Water footprint outcomes and policy relevance change with scale considered: evidence from California. Water Resour Manag 28(11):3637–3649

Ghassemi F, White I (2006) Inter-basin water transfer: case studies from Australia, United States, Canada. Cambridge University Press, China and India

Gleick PH (2003) Global freshwater resources: soft-path solutions for the 21st century. Science 302(5650):1524–1528

Gu X, Liao Z, Zhang G, Xie J, Zhang J (2017) Modelling the effects of water diversion and combined sewer overflow on urban inland river quality. Environ Sci Pollut Res 24(26):21038–21049

Gualtieri AG (2009) Legal implications of trade in real and virtual water resources. In: Water Law for the Twenty-First Century. Routledge, pp 79–100

He C, He X, Fu L (2010) China's South-to-north water transfer project: is it needed? Geogr Compass 4(9):1312–1323

Hoekstra AY, Chapagain AK (2006) Water footprints of nations: water use by people as a function of their consumption pattern. Water Resour Manag 21:35–48

Hoekstra AY, Mekonnen MM (2012) The water footprint of humanity. Proc Natl Acad Sci 109(9):3232–3237

Huang J, Gao J, Zhang Y, Xu Y (2015) Modeling impacts of water transfers on alleviation of phytoplankton aggregation in Lake Taihu. J Hydroinformatics 17(1):149

Huang J, Yan R, Gao J, Zhang Z, Qi L (2016) Modeling the impacts of water transfer on water transport pattern in Lake Chao, China. Ecol Eng 95:271–279

Huang T, Pang Z (2010) Changes in groundwater induced by water diversion in the Lower Tarim River, Xinjiang Uygur, NW China: evidence from environmental isotopes and water chemistry. J Hydrol 387:188–201

Islar M, Boda C (2014) Political ecology of inter-basin water transfers in Turkish water governance. Ecol Soc 19(4):15

Konar M, Caylor KK (2013) Virtual water trade and development in Africa. Hydrol Earth Syst Sci 17(10):3969–3982

Kumar MD, Singh OP (2005) Virtual Water in global food and water policy making: Is there a need for rethinking? Water Resour Manag 19(6):759–789

Lenzen M, Moran D, Bhaduri A, Kanemoto K, Bekchanov M, Geschke A, Foran B (2013) International trade of scarce water. Ecol Econ 94:78–85

Li L, Zhang L, Xia J, Gippel CJ, Wang R, Zeng S (2015) Implications of modelled climate and land cover changes on runoff in the middle route of the south to north water transfer project in China. Water Resour Manag 29:2563–2579

Li Y, Acharya K, Yu Z (2011) Modeling impacts of Yangtze River water transfer on water ages in Lake Taihu, China. Ecol Eng 37(2):325–334

Li Y, Tang C, Wang C, Tian W, Pan B, Hua L, Lau J, Yu Z, Acharya K (2013) Assessing and modeling impacts of different inter-basin water transfer routes on Lake Taihu and the Yangtze River, China. Ecol Eng 60:399–413

Liu C, Zheng H (2002) South-to-north water transfer schemes for China. Int J Water Resour Dev 18:453–471

Liu J, Wang Y, Yu Z, Cao X, Tian L, Sun S, Wu P (2017) A comprehensive analysis of blue water scarcity from the production, consumption, and water transfer perspectives. Ecol Indic 72:870–880

Liu J, Zang C, Tian S, Liu J, Yang H, Jia S, You L, Liu B, Zhang M (2013) Water conservancy projects in China: achievements, challenges and way forward. Glob Environ Chang 23(3):633–643

Liu J, Zehnder AJB, Yang H (2009) Global consumptive water use for crop production: the importance of green water and virtual water. Water Resour Res 45(5):W05428

Ma Y, Li XY, Wilson M, Wu XC, Smith A, Wu J (2016) Water loss by evaporation from China's south-north water transfer project. Ecol Eng 95:206–215

Magee D (2011) Moving the river? China's south–north water transfer project. In: Engineering Earth. Springer, pp. 1499–1514

Manshadi HD, Niksokhan MH, Ardestani M (2015) A quantity-quality model for inter-basin water transfer system using game theoretic and virtual water approaches. Water Resour Manag 29(13):4573–4588

Mekonnen MM, Hoekstra AY (2011) The green, blue and grey water footprint of crops and derived crop products. Hydrol Earth Syst Sci 15(5):1577–1600

Merrett S (2003) Virtual water and the Kyoto consensus: a water forum contribution. Water Int 28(4):540–542

Novo P, Garrido A, Varela-Ortega C (2009) Are virtual water flows in Spanish grain trade consistent with relative water scarcity? Ecol Econ 68:1454–1464

Peng A, Peng Y, Zhou H, Zhang C (2014) Multi-reservoir joint operating rule in inter-basin water transfer-supply project. Sci China Technol Sci 58(1):123–137

Pittock J, Meng J, Geiger M, Chapagain KA (2009) Inter-basin water transfers and water scarcity in a changing world—a solution or a pipedream? WWF Germany. http://hydropower.inel.gov/turbines/pdfs/doeid-13741

Porkka M (2011) The role of virtual water trade in physical water scarcity: Case Central Asia. Master thesis, Aalto University

Qadir M, Sharma BR, Bruggeman A, Choukr-Allah R, Karajeh F (2007) Non-conventional water resources and opportunities for water augmentation to achieve food security in water scarce countries. Agric Water Manag 87(1):2–22

Reimer JJ (2012) On the economics of virtual water trade. Ecol Econ 75:135–139

Ridoutt BG, Juliano P, Sanguansri P, Sellahewa J (2009) Consumptive water use associated with food waste: case study of fresh mango in Australia. Hydrol Earth Syst Sci Discuss 6(4):5085–5114

Roson R, Sartori M (2010) Water Scarcity and Virtual Water Trade in the Mediterranean

Roth D, Warner J (2007) Virtual water: virtuous impact? the unsteady state of virtual water. Agric Hum Values 25(2):257–270

Shao X, Wang H, Wang Z (2003) Interbasin transfer projects and their implications: a China case study. Int J River Basin Manag 1(1):5–14

Shrestha S, Pandey VP, Chanamai C, Ghosh DK (2013) Green, blue and grey water footprints of primary crops production in Nepal. Water Resour Manag 27(15):5223–5243

Shrestha S, Shrestha M, Babel MS (2015) Assessment of climate change impact on water diversion strategies of melamchi water supply project in Nepal. Theor Appl Climatol 128(1–2):311–323

Sun SK, Wang YB, Engel BA, Wu P (2016) Effects of virtual water flow on regional water resources stress: a case study of grain in China. Sci Total Environ 550:871–879

Tamea S, Carr JA, Laio F, Ridolfi L (2014) Drivers of the virtual water trade. Water Resour Res 50:17–28

The USGS Water Science School (2016) How much water is there on, in, and above the Earth? https://water.usgs.gov/edu/earthhowmuch.html. Accessed 2 December 2016

Vos J, Boelens R (2016) The politics and consequences of virtual water export. Eating, Drinking: Surviving pp 31–41

Wichelns D (2001) The role of virtual water in efforts to achieve food security and other national goals, an example from Egypt. Agric Water Manag 49(2):131–151

Wichelns D (2015) Virtual water and water footprints do not provide helpful insight regarding international trade or water scarcity. Ecol Indic 52:277–283

Yang H, Zehnder A (2001) China's regional water scarcity and implications for grain supply and trade. Environ Plan A 33(1):79–95

Yang H, Zehnder A (2002) Water scarcity and food import: a case study for southern Mediterranean countries. World Dev 30:1413–1430

Yang H, Zehnder A (2007) Virtual water: an unfolding concept in integrated water resources management. Water Resour Res 43(12):W12301

Yang SQ, Liu PW (2010) Strategy of water pollution prevention in Taihu Lake and its effects analysis. J Gt Lakes Res 36(1):150–158

Yang Y, Yin L, Zhang Q (2015) Quantity versus quality in China's south-to-north water diversion project: a system dynamics analysis. Water 7(5):2142–2160

Yawson DO, Mulholland B, Tom B, Sushil M, White P (2013) Food security in a water-scarce world: making virtual water compatible with crop water use and food trade. Manag, Econ Eng Agric Rural Dev 13(2):431–443

Yu Y, Hubacek K, Feng K, Guan D (2010) Assessing regional and global water footprints for the UK. Ecol Econ 69(5):1140–1147

Zeng Z, Liu J, Savenije HHG (2013) A simple approach to assess water scarcity integrating water quantity and quality. Ecol Indic 34:441–449

Zhan A, Zhang L, Xia Z, Ni P, Xiong W, Chen Y, Haffner GD, MacIsaac HJ (2015) Water diversions facilitate spread of non-native species. Biol Invasions 17(11):3073–3080

Zhang E, Yin X, Xu Z, Yang Z (2017a) Bottom-up quantification of inter-basin water transfer vulnerability to climate change. Ecol Indic 92:195–206

Zhang Y, Zhang J, Wang C, Cao J, Liu Z, Wang L (2017b) China and trans-pacific partnership agreement countries: estimation of the virtual water trade of agricultural products. J Clean Prod 140:1493–1503

Zhang H, Ma S, Zhang X, Wang Y (2010) Analysis of Tianjin Virtual Water Trade Based on Input-Output Model. International Conference on System Science and Engineering: 21–25

Zhang L, Li S, Loáiciga HA, Zhuang Y, Du Y (2015) Opportunities and challenges of interbasin water transfers: a literature review with bibliometric analysis. Scientometrics 105:279–294

Zhao X, Chen B, Yang ZF (2009) National water footprint in an input–output framework—a case study of China 2002. Ecol Model 220(2):245–253

Zhao X, Liu J, Liu Q, Tillotson MR, Guan D, Hubacek K (2015a) Physical and virtual water transfers for regional water stress alleviation in China. Proc Natl Acad Sci 112(4):1031–1035

Zhao ZY, Zuo J, Zillante G (2015b) Transformation of water resource management: a case study of the South-to-North Water Diversion Project. J Clean Prod 163:136–145

Zhuang W (2016) Eco-environmental impact of inter-basin water transfer projects: a review. Environ Sci Pollut Res 23(13):12867–12879

Chapter 3
Pattern of Physical and Virtual Water Flows: The Impact to Water Quantity Stress Among China's Provinces

3.1 Introduction

The difference in water distribution around the world has generated a gap between water for needs and water resources availability. Most regions such as the Middle East, some part of Asia, North Africa have been suffering from it. They either externalise their water supply to meet demands and in return release water stress within the supplying territories (through virtual water), or bring water to the place of use (physical water diversion). Similarly, China has been facing severe water shortage problems due to the rapid socio-economic development, climate change, rapid urbanisation, and industrialisation (Cai et al. 2017). Several regions in China face water scarcity issues. The North China Plain experiences the highest pression on its water endowment. It is the main breadbasket of China, it contains approximately 25×10^6 ha of cultivated land (i.e., a quarter of the Chinese total), yet only possesses ten per cent of the nation's water resources, or 1 / 20th of the world per capita average (Zhang 2009). The water shortage restricts the sustainable development of China's society and economy. Sixty-six per cent of China's cropland is in the north, but only 25% of China's total water is available (Cheng and Song 2009). A large amount of the Chinese virtual water flow occurs in form of agricultural products, leading to increasing water stress in the water-exporting provinces. China relies on physical water transfer and virtual water to support their demands.

As such, some have assessed virtual water flows between China and the world (Chen et al. 2017; Zhang et al. 2017; Shi et al. 2014). They argued that at a global scale China was a net virtual water importer for trade in agricultural commodities and a net virtual water exporter for all sectors. The geography of the study area with its provinces and regions is represented in Fig. 3.1: China's provinces. Note: the provinces where data are not available are shown in white. This study focuses only on the mainland regions where data were available. The bold line separates regions, and the shade is made for a clear distinction among provinces. The following sections are divided into three parts; first the pattern of physical water transfer within China,

© The Author(s), under exclusive license to Springer Nature Singapore Pte Ltd. 2020
Y. Li et al., *Addressing the Uneven Distribution of Water Quantity and Quality Endowment*, SpringerBriefs in Water Science and Technology, https://doi.org/10.1007/978-981-13-9163-7_3

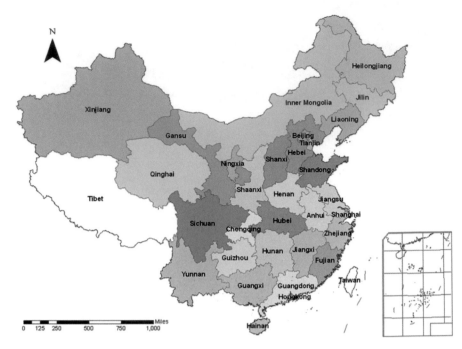

Fig. 3.1 China's provinces. *Note* the provinces where data are not available are shown in white

secondly virtual water flow among Chinese provinces, and finally the impacts of both physical and virtual water on water stress.

3.2 Physical Water Pattern Within China

The growing population's demands to sustain their lifestyle has been the core of many issues on water resources within China, particularly, on water stress-induced water scarcity and environment. As such, regardless of their water endowment, most of China's regions have been involved in physical water diversion. Depending on the amount of water which has been exchanged among provinces and the water stress due to production and consumption, the pattern of physical water transfer within China can be obtained.

From Fig. 3.2, the provinces relying on physical water transfer and the level of water stress have been represented. In 2007, most of the provinces in the northern part were in water-stressed condition, Shanghai which is a coastal area also presents water-stressed condition. While the others are in moderate or no stress condition. Consequently, they relied on water supply outside their territory. Beijing, Tianjin, and Shanghai which are densely populated megacities have used diverted water as supply for their needs. Moreover, on a national scale, physical water flows by water

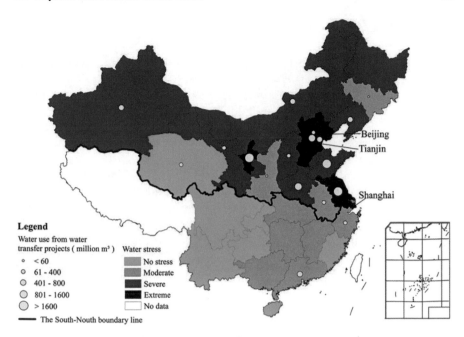

Fig. 3.2 Water stress evaluation of China's provinces. The colour coding makes the distinction among the different levels of water stress. The size of the dots reflects the amount of physical water transfer (Zhao et al. 2015a)

Table 3.1 Virtual water imports and exports by region (10^9m^3) (Ma et al. 2006)

	From other region within China			From outside China			Overall net virtual water import	Net virtual water import per capita
	Gross virtual water import from north	Gross virtual water import from south	Net virtual water import	Gross virtual water import	Gross virtual water export	Net virtual water import		
North	–	−51.6	−51.6	8.4	16.2	−7.8	−59.4	−102
South	51.6	–	51.6	19.7	2.7	17.0	68.6	104
National total	–	–	0	28.1	18.9	9.2	9.2	7

transfer projects amounted to 26.3 Gm^3, accounting for 4.5% of national water supply in China. Table 3.1 globalises for each region water use both within their territory and outside.

3.3 Virtual Water Flows Pattern Within China

Virtual water has been considered as a promising tool to relieve water stress within regions/countries. China, which is the world's most populated country has also considered it. There is a number of studies on virtual water flow within China (Zhao et al. 2015a; Ma et al. 2006; Cai et al. 2017). They differ from one another based on the consideration of commodities and services and areas. Some focus only on agricultural commodities whereas others take into account livestock and industries (Xinchun et al. 2011; Liu et al. 2013).

This study present virtual water flows within China. The baseline is noted for each result. The quantification of Virtual water within China has been obtained from FAO databases CropWat (http://www.sdnbd.org/sdi/issues/agriculture/database/CROPWAT.htm), ClimWat (http://www.fao.org/landandwater/aglw/climwat.stm), and FAOSTAT (http://faostat.fao.org/). For China and foreign countries, data have been taken from Chapagain and Hoekstra (2004) and Hoekstra and Hung (2002). The digraph (Fig. 3.3) depicts the flows of virtual water in agricultural and livestock products among eight sub-areas within China. From this digraph (Fig. 3.3), at a regional level, the southern part is the principal exporter of virtual water. Around 52 billion m^3 (Table 3.1) of virtual water flowed from the northern part to the southern part in 1999 (Ma et al. 2006). The most net importers areas of virtual water (among regions within China) are located in north central, southeast and south-central where a huge amount of virtual water has amounted to 14.1, 25.8, 25.8 (10^9m^{3}yr.$^{-1}$) respectively. Table 3.1 summarises the virtual water import and export between North and South and outside of China. More specifically, virtual water flows from the major exporters to the major importers among Chinese provinces have been depicted in Fig. 3.3 and Fig. 3.4 respectively.

Figure 3.4 obtained from MRIO model also shows that in 2007, the most importing areas of virtual water (trade in all sectors) are located in the eastern part of the country specifically along the coast. Those regions which are the most developed and populated have imported commodities from the northern regions through virtual water. The top five net importers are Shandong, Shanghai, Guangdong, Zhejiang, and Tianjin with the transferred virtual water amounting to 14.5, 14.3, 12.4, 10.6 and 9.2 (expressed in Gm^3) respectively. Meanwhile, Xinjiang, Heilongjiang, Inner Mongolia which are less developed and water scarce areas became the major exporters of virtual water to the rest of China's regions. The amounts of virtual water flow from the latter regions are 24.4, 12.8, and 9.7 (expressed in Gm^3) respectively. At last the total volume of virtual water flows was 201Gm^3 in 2007.

3.3.1 Virtual Water Flows Per Sectors Within China

As mentioned in the previous part virtual water includes commodities and services. As such, based on some authors' works (Zhao et al. 2009; Chapagain and Hoekstra

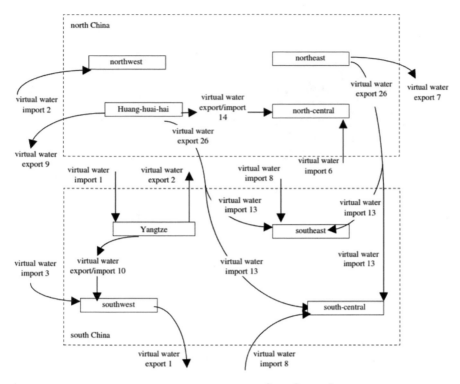

Fig. 3.3 Virtual water flow among regions in China ($10^9 \times m^3 \times yr.^{-1}$) (Ma et al. 2006)

2008) and the availability of data related to it, commodities and services have been classified into subgroups. The main sectors of virtual water import and export within China are depicted in Figs. 3.5 and 3.6.

From Figs. 3.5 and 3.6, agricultural sector made the largest contribution to virtual water flows. Virtual water imports from the agricultural and chemical sectors of the major importers which included Guangdong, Shanghai, and Beijing represented more than 75% of their total virtual water imports (see Fig. 3.4). These two sectors represent alone the greatest contribution to internal water consumption. This indicates that imported virtual water includes a large number of water-intensive and heavily-polluting products originate from exporting provinces. In addition, the three main sectors namely the agricultural, chemical industry and petroleum processing, and coking industry sectors are the principal contributors to virtual water flows. Virtual water flows from these three sectors accounted for 66.8%, 7.1% and 6.2% of the total virtual water flows, respectively.

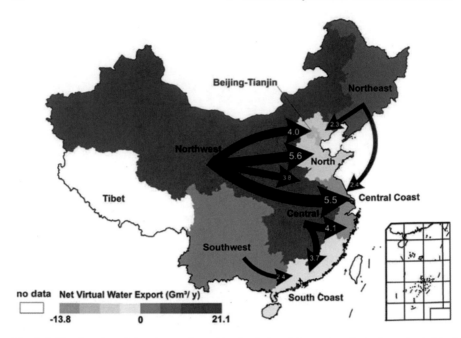

Fig. 3.4 Net direction of virtual water flows. For the clarity of the graph, only the largest net virtual water flows are shown (> 2Gm^{3}yr.$^{-1}$) (Zhao et al. 2015a)

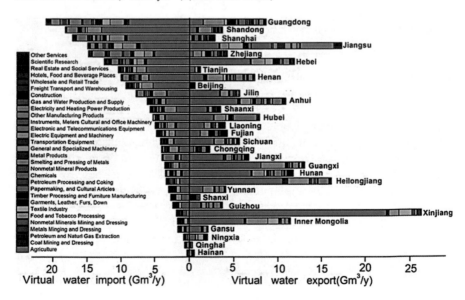

Fig. 3.5 Virtual water flows import and export for each province in different sectors (Zhao et al. 2015a)

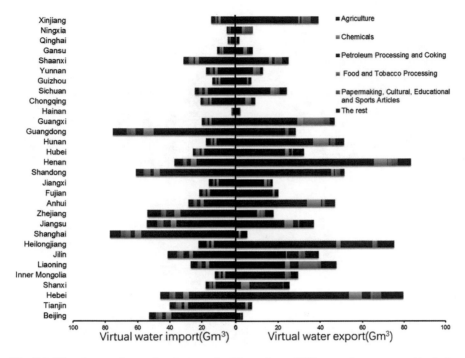

Fig. 3.6 Virtual water flows of each sector for 30 provinces. Different colours represent trade in domestic final goods by sector (Cai et al. 2017)

3.4 Impacts on Water Stress

We calculated water stress as the ratio of water withdrawal to renewable freshwater resources within a province (Zhao et al. 2015a). The distinction between actual water stress (WSI) and hypothetical water stress (WSI*) has been made. The latter would refer to the hypothetical water stress on the local hydro ecosystem if the importing province were not to have physical and virtual water inflows available to it (i.e., it would be required to withdraw all required water from local sources) (Zhao et al. 2015a). Hence, the difference between WSI* and WSI represents the contribution of net virtual and physical water flows in terms of increasing or ameliorating water stress.

Among provinces which are in water-stressed condition, 12 provinces benefited from both physical and net virtual water imports (Fig. 3.7) (WSI* > WSI), amounted to 80 Gm³ and 5 Gm³ respectively. Beijing, Tianjin, Shanghai are the most beneficiaries from it. Water stress index in the perspective of final consumption (WSI*) in these provinces was enhanced, but the water stress in the perspective of local water resources (WSI) was still significant (above the moderate threshold). Water stress of other provinces remained at the same level in spite of their physical and net virtual water imports as to those which were already in water-stressed condition their situ-

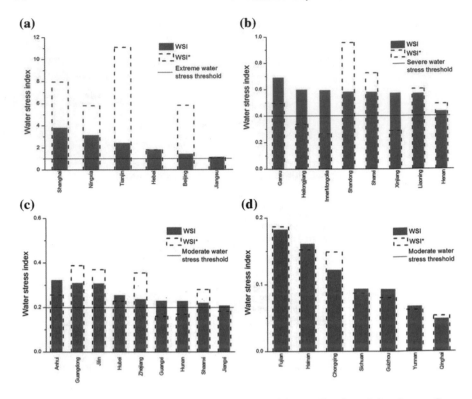

Fig. 3.7 Virtual water balance per economic regions and the net direction of virtual water flows. (> 2Gm^{3}yr.$^{-1}$) (Zhao et al. 2015a)

ation worsened (WSI* < WSI). Heilongjiang, Inner Mongolia, Xinjiang, Guangxi, Hunan, and Jiangxi experienced water stress induced by the consumption of other provinces (Fig. 3.7).

3.5 Conclusion

In summary, China's provinces are at different levels of water endowment. When producing goods and services such as agricultural products and industrial products, China's water-scarce provinces exacerbate water stress on their water endowment. Physical water transfer projects and virtual water flows create a re-allocation of water within China's boundaries. The pattern of virtual water showed that huge amount of virtual water moved from the Northern provinces to the Eastern provinces, creating a king of outsourcing water supply. Water stress as an indicator revealed that about 12 provinces significantly lowered their water stress level by both physical and virtual water flows. However, attention should be paid to the management of trade policy.

References

Cai B, Wang C, Zhang B (2017) Worse than imagined: unidentified virtual water flows in China. J Environ Manag 196:681–691

Chapagain AK, Hoekstra AY (2004) Water footprints of nations

Chapagain AK, Hoekstra AY (2008) The global component of freshwater demand and supply: an assessment of virtual water flows between nations as a result of trade in agricultural and industrial products. Water Int 33:19–32

Chen W, Wu S, Lei Y, Li S (2017) Virtual water export and import in China's foreign trade: a quantification using input-output tables of China from 2000 to 2012. Resour Conserv Recycl 132:278–290

Cheng S, Song H (2009) Conservation buffer systems for water quality security in south to north water transfer project in China: an approach review. Front For China 4:394–401

Hoekstra AY, Hung PQ (2002) Virtual water trade. A quantification of virtual water flows between nations in relation to international crop trade. Value of water research report series 1, The Netherlands

Liu J, Wu P, Wang Y, Zhao X, Sun S, Zhang X (2013) Analysis of virtual water flows related to crop transfer and its effects on local water resources in Hetao irrigation district, China, from 1960 to 2008. J Food Agric Environ 11(1):682–686

Ma J, Hoekstra AY, Wang H, Chapagain AK, Wang D (2006) Virtual versus real water transfers within China. Philosophical transactions of the Royal Society of London. Ser B, Biol Sci 361:835–842

Shi J, Liu J, Pinter L (2014) Recent evolution of China's virtual water trade: analysis of selected crops and considerations for policy. Hydrol Earth Syst Sci 18:1349–1357

Xinchun C, Pute W, Yubao W, Xining Z, Sha L (2011) Application of virtual water trade theory in interregional grain allocation and transportation in China. African J Biotechnol 10(80):18463–18471

Zhang Q (2009) The south-to-north water transfer project of China: environmental implications and monitoring strategy. J American Water Resour Assoc 45:1238–1247

Zhang Y, Zhang J, Wang C, Cao J, Liu Z, Wang L (2017) China and trans-pacific partnership agreement countries: estimation of the virtual water trade of agricultural products. J Clean Prod 140:1493–1503

Zhao X, Chen B, Yang ZF (2009) National water footprint in an input-output framework—a case study of China 2002. Ecol Model 220:245–253

Zhao X, Liu J, Liu Q, Tillotson MR, Guan D, Hubacek K (2015a) Physical and virtual water transfers for regional water stress alleviation in China. Proc Natl Acad Sci USA 112:1031–1035

Zhao ZY, Zuo J, Zillante G (2015b) Transformation of water resource management: a case study of the south-to-north water diversion project. J Clean Prod 163:136–145

Chapter 4
Physical Water Transfer and Its Impact on Water Quality: The Case of Yangtze River Diversions

4.1 Introduction

Water transfer engineering is an essential approach for lake restoration. It has been widely used for water supply and energy production around the world. Besides, it has been successfully applied in several water bodies for different purposes, namely, for accelerating water exchange, diluting polluted water, improving water quality, and mitigating eutrophication issues. These can be achieved by transferring large volumes of water from a relatively clean source to a severely polluted water body. Germany is one of many countries which has built an engineering channel to divert water for ecological restoration purpose (the Bavaria Water transfer Project) (Zhuang 2016). Similarly, in China, there are many different water transfers depending on the purposes (water supply, hydropower, environmental restoration), for example, the SNWTP, inter-basin water transfer and so on. This chapter focuses on two specific water transfer projects in China. Those are water transfer into the Lake Taihu and that into the Lake Chao. The aim of those water transfers is for flushing out pollutants into the lake to reach China's water quality standard.

This chapter is divided into two parts, the first is dedicated to Lake Taihu and the second to Lake Chao. Each part is presented as follow: presentation of the study area, then the method, and finally results and discussion.

4.2 Physical Water Transfer: The Case of Lake Taihu

Lake Taihu, one of the largest freshwater lake in China with water surface area of about 2338 km^2, has been facing serious issues such as eutrophication. As such, the growing of aquatic plants, specifically algae is increasing due to the discharge of excess pollutants from human activities. Consequently, the water quality and

Y. Li et al., *Addressing the Uneven Distribution of Water Quantity and Quality Endowment*, SpringerBriefs in Water Science and Technology, https://doi.org/10.1007/978-981-13-9163-7_4

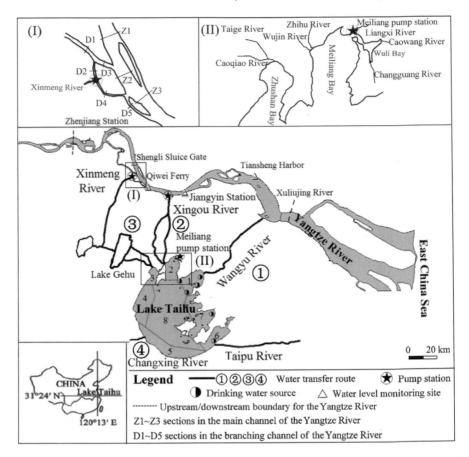

Fig. 4.1 The location of Lake Taihu, Yangtze River watershed and the main tributaries. The bold lines and digitals represent different water transfer routes. (I) the details about the topography around route Three in the Yangtze River; Z1–Z3 and D1–D5 represent sections in the main and branching channel of the Yangtze River, respectively (Li et al. 2013b)

dimensions of the lake may be threatened to offset by decreasing water availability for any purposes. The water transfer consists of four routes to transfer clean water from source (clean here means less polluted) into the lake to flush out pollutants and, therefore ameliorate its quality. The different routes of water transfer from the Yangtze River into the lake Taihu are shown in Fig. 4.1. As shown in the previous Fig. 4.1 the eight sub-areas of Lake Taihu are: (1) Gonghu Bay, (2) Meiliang Bay, (3) Zhushan Bay, (4) Northwest Zone, (5) Southwest Zone, (6) Dongtaihu Bay, (7) East Epigeal Zone, (8) Central Zone.

Route one transfers freshwater from the Yangtze River into Lake Taihu via the Wangyu River and discharges water through the Taipu River. The aim of route 2 is to enhance water exchange in Meiliang Bay in the northern part of the lake by adding two pump stations named Meiliang and Xingou around Meiliang Bay (Fig. 4.1), as to routes 3 and 4, they are for conveying freshwater from the Yangtze River to the northwest, and southwest lake region via the Xinmeng River and the Changxing River respectively.

4.2.1 Numerical Model

The three-dimensional hydrodynamic EFDC model was used to simulate water levels, currents, tracer concentrations, and water ages. Lagrange particle tracking which determines the path of water parcels from the different water transfer routes highlights the results.

4.2.1.1 Water Age

The concept of water age is used to characterise properties of the lake. It is defined as the time that a particle takes from one point considered at reference time, $t = 0$, to reach a given second point. The mean water age is calculated as follows:

$$a(t, \vec{x}) = \frac{\alpha(t, \vec{x})}{c(t, \vec{x})} \tag{4.1}$$

Where α and c are respectively the age concentration and the tracer concentration. The following equations give their expressions.

$$\frac{\partial c(t, \vec{x})}{\partial t} + \nabla\left(uc(t, \vec{x}) - k\nabla c(t, \vec{x})\right) = 0 \tag{4.2}$$

$$\frac{\partial \alpha(t, \vec{x})}{\partial t} + \nabla\left(u\alpha(t, \vec{x}) - k\nabla \alpha(t, \vec{x})\right) = c(t, \vec{x}) \tag{4.3}$$

In these equations, u is the velocity field, k is the diffusivity tensor, t is time, and $\vec{x}$ is coordinate.

4.2.2 Results and Discussion

The concept of particles tracking is to release particles into a lake (where any flow exists) and follow their movement within given boundaries, thereby, the displacement from the initial position to the final would be used to draw patterns. In this case, the particles initially marked with red triangles have been released into the lake through specific inlets: Zhushan Bay, Gonghu Bay, and the middle area of Meiliang. Then,

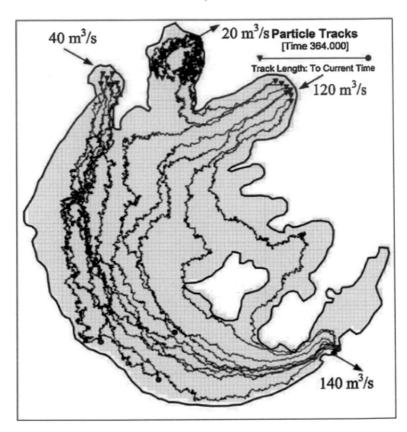

Fig. 4.2 Pathway of water parcel from the Route One (120 m³/s), Route Two (20 m³/s) and Route Three (40 m³/s) (the red triangle and blue circles represent the released and end position, respectively) (Li et al. 2013b)

the final positions of the particles within the lake have been marked with red circles to bring out the water parcel pathways from the water transfer routes Fig. 4.2.

It should be noted that these results have been obtained for the optimal water transfer scheme of the routes one to three (see Li et al. 2013a), and the model scenarios in different wind conditions are shown in Table 4.1. From that Fig. 4.2 the different paths appeared. The particles entering through route three mainly covered Zhushan Bay, the Northwest Zone, parts of the Southwest Zone, and small portions of the Central Zone. The paths of particles from route One illustrated that this route was beneficial to improve water exchanges in Gonghu Bay, the East Epigeal zone, Dongtaihu Bay, and the major Central Zone. The particles in Meiliang Bay were taken out by the Meiliang pump station, which contributed to the water exchange in Meiliang Bay.

Table 4.1 Model simulation scenarios (Li et al. 2013a)

Model scenarios	Winds	Flow discharge (unit: m³/s)			
		Wangyu river (inflow)	Taipu river (outflow)	Meilang pump station (outflow)	Xingou pump station (outflow)
Group one	No wind Wind speed 5 m/s, wind directions are N, NE, E, SE, S, SW, W, NW, respectively	From 50 to 200 increased by 5	From 50 to 200 increased by 5	–	–
Group two	Wind speed 5 m/s, wind directions are N, NE, E, SE, S, SW, W, NW, respectively	Optimal inflow obtained from group one	Keep water balance: outflow from Taipu River = inflow from meilang pump station	From 5 to 50 increased by 5	–
Group three	Wind speed 5 m/s, wind direction is SE	Optimal inflow obtained from group one	Keep water balance: outflow from Taipu River = inflow from Wangyu River – outflow from Meilang pump station- outflow from Xingou pump station	Optimal inflow obtained from group two	20

4.2.2.1 Effect of Wind and Inflow Rate in Water Age

Water Age in the Whole Lake

We considered parameters such as total nitrogen, total phosphorene, and the volume of water in each lake region, to evaluate water age in Taihu Lake. The results showed that the more water is transferred from Yangtze River to the lake, the lower average water age in the lake is (Fig. 4.3a). In addition, water age is strongly related to the quantity of the transferred water. Considering east wind scenario as an example, water age significantly decreased from 300 to 174 days as inflows increased from 50 to 200 m³/s.The slight changes observed in inflow rate leads to high changes

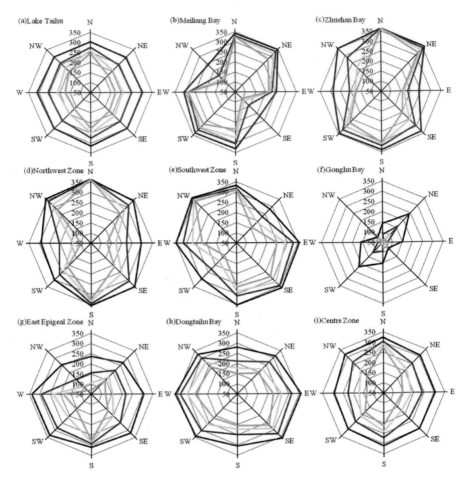

Fig. 4.3 Water age on Julian day 365 in eight lake zones under eight different wind direction and four flow discharges (i.e. 50 m³/s-black line, 100 m³/s-red line, 150 m³/s-blue line, 200 m³/s-yellow line). **a** the entire lake considering the weights of each lake region; **b** Meiliang Bay; **c** Zhushan Bay; **d** Northwest Zone; **e** Southwest Zone; **f** Gonghu Bay; **g** East Epigeal Zone; **h** Dongtaihu Bay; **i** Central Zone (Li et al. 2013a)

in average water age in the lake. For example, under the southeast wind, when the transferred inflow increase from 50 to 100 m³/s, 100 to 150 m³/s the averaged water age in the lake decreased subsequently by 26.14 and 16.20% respectively.

The pattern of water age in different wind directions was similar apart from that in the north and south wind. North and south wind had a greater impact on water age. Under these previous wind directions, water age was more than 50 days. However, changes in water ages were less than 13 days under most other wind directions. Besides, for specific inflows from Wangyu River, water age in the north wind scenario and the south wind was the highest. North and south wind do not have a great contribution to enhance the efficacy of the water transfer. Water age was much influenced by the flow rates rather than wind directions. The variation of water age evolved in the same direction as the flow rates (Fig. 4.3a). Moreover, the average water age in the lake was equal to 300 days with 50 m^3/s inflow in all wind directions. However, the water age ranged from 150 to 280 days in the higher flow rate of 200 m^3/s. Wind, especially its direction was a significant parameter and a driving factor influencing water age in the lake.

Water Age in Specific Lake Regions

Because each part of the lake presents a different level of pollution, and it is the water source for different purposes e.g., drinking, inflows from water transfer and winds in different areas of the lake were investigated, since both can impede water age in the lake. The results showed that the water age in each area of the lake was different due to wind direction and transferred inflow (Fig. 4.3). In Meiliang Bay, the highest contribution on water age caused by winds was almost twice as much as that of the transferred water (Fig. 4.3b). Northwest and southeast wind directions play a great role in enhancing water age. In northwesterly wind, water age reached 198 days when the transferred inflow increased from 50 to 200 m^3/s. Besides, water age in particular areas of the lake including Zhushan Bay, Northwest and Southwest zones is more influenced by wind directions than water inflow. The ratio of wind versus inflow contributions on water age were about 1.2, 1.4, and 1.3 in Zhushan Bay, Northwest and Southwest zones, respectively (Fig. 4.3c–e). On the contrary, for Gonghu Bay, East Epigeal Zone, and Central Zone, the contribution of water inflow was about 60% and that of wind was about 40% (Fig. 4.3f, g, i). For the area located at the outlet of the water transfer (Dong- Taihu Bay), the effect of the transferred water and wind play an equal role on water age (Fig. 4.3h).

Results also suggested that each lake region had its favorite wind direction to improve water transfer and enhance water exchange capacity. The lowest water age in Meiliang Bay occurred due to the southeast and northwest winds, while northeast wind impeded this process. For example, when the transferred inflow from Yangtze River was 150 m^3/s, the averaged water ages in Meiliang Bay were 102 and 198 days under the southeasterly and northwesterly wind, much smaller than that under the northeasterly wind (Fig. 4.3b). Water age in some polluted areas such as Zhushan Bay, Northwest Zone, and Southeast Zone is lower under easterly, westerly and north-easterly wind directions, respectively, while north and south winds, they worsened water age in Zhushan Bay and Northwest Zone. Water ages ranged between 160 and 240, 98 and 290, 157 and 284 days for desirable winds with increasing inflows for Zhushan Bay, Northwest Zone and Southwest Zone, respectively.

Generally, the north and south wind directions were not helpful to improve water age regardless of the amount of water inflow. The youngest water age was in Gonghu Bay, and it was about 80 days (Fig. 4.3f). North wind had a great effect in enhancing water exchange in East Epigeal Zone and Dongtaihu Bay (Fig. 4.3g, h). The water age for Central Zone under desirable wind directions (easterly and westerly winds) was about 132–300 days when transferred inflows from Wangyu River varied from 50 to 200 m^3/s (Fig. 4.3i). In general, water ages were the smallest in regions close to Wangyu River and the route of the original Yangtze River transfer. However, there were no significant changes in water ages in the heavily polluted lake regions such as Meiliang Bay and Zhushan Bay due to the original Yangtze River transfer.

4.2.2.2 Effect of Different Routes in Water Age

On the perspective of water age, the results showed that the contribution of route one and three was similar, and they led to reducing water age in the lake Taihu of about 235 days Fig. 4.4a, b. The pattern of water age across the lake was different because of the relative position of the inflow. The part of the lake which is the main source of drinking water (Gonghu Bay) was the top beneficiary with the youngest water age of about 18 days. Whilst the most polluted areas include Zhushan Bay, Northwest Zone, Southwest Zone, and Meiliang Bay benefited the least (Fig. 4.4a). On the contrary, route three brought much more improvement for water exchange in the polluted zones except Meiliang Bay and some parts of the Southwest Zone (Fig. 4.4b). Water age of the eastern region is a little longer corresponding to 300 days and, thereby, leading to weak water exchange in the concerned area.

The effect of the two routes on improving water exchange in the lake, specifically, in Meiliang Bay might not be done without proper combinations. Operate properly (see Group three, Table 1 in Li et al. 2013b) routes one, three, and four would be a significant contributor to enhancing water exchange into the lake. In that case, acting together route one and two led to decreasing water age in Meiliang Bay from 286 to 133 days (Fig. 3.4a, c). Similarly, using route two and three reduces water age in Meiliang Bay by 44.78% compared to the use of route three only (Fig. 4.4b, d).

Two cases have been investigated, the algae bloom and the non-algae bloom periods. The results obtained during the two periods have been highlighted in Fig. 4.5. Some conditions have been taken into consideration to get the optimal scheme of the water transfer into the lake (see Li et al. 2013b).

As such, in non-algae bloom period, the best operation scheme for inflow from routes one, three, and four were of 90, 70 and 20 m^3/s, respectively, while the transferred water would flow out through route two at 40 m^3/s, respectively. Correspondingly, water age in the entire lake was about 160 days and 98.4% of the area had been exchanged with relatively low economic input. In the areas such as Meiliang Bay, Zhushan Bay, Northwest Zone and Southwest Zone which are polluted, water age was about 10, 112, and 212 days respectively (Fig. 4.5a). Although, the best operation scheme in the algae bloom period was to have inflow from routes One, Three, and Four of 80, 100, and 20 m^3/s respectively, while the outflow of route two was 70 m^3/s. Water age in this situation was 10–20 days younger than in the previous situation, particularly, in the heavily polluted lake regions (Fig. 4.5b).

In summary, each route has played a role in the water quality improvement of the lake. Inflow from route one has enhanced water exchange in the entire lake by decreasing water age in many parts of the lake. Route two has a significant effect

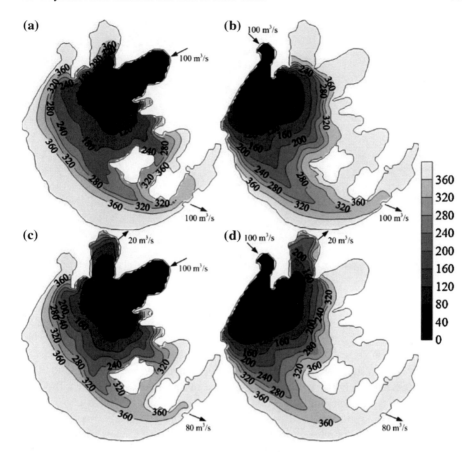

Fig. 4.4 Spatial distribution of water age in Lake Taihu with different water transfer routes. **a** Route one (100 m³/s); **b** Route three (100 m³/s); **c** Route One (100 m³/s) and Route two (20 m³/s); **d** Route Three (100 m³/s) and Route Two (20 m³/s) (Li et al. 2013b)

on water exchange in Meiliang Bay which was a blind zone for Route one and route three. Route three has a high impact on Zhushan Bay and the Northwest Zone where eutrophication levels and algae blooms frequently arose, as to route 4 operated mostly in the Southwest Zone where the deficiency of the routes one and three to solve eutrophication issue have been observed.

4.3 Physical Water Transfer: The Case of Lake Chao

Lake Chao watershed is formed by six inflows (Hangbu River, Baishitian River, Zhao River, Zhegao River, Nanfei River and Pai River) and one outflow (Yuxi River) connecting with Yangtze River (Huang et al. 2016). It covers an area of 13,555 km²

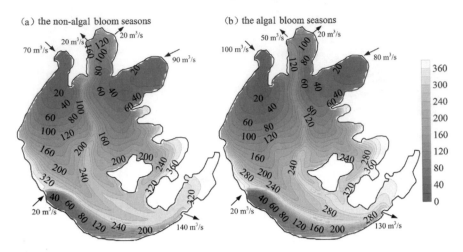

Fig. 4.5 Spatial distribution of water age in Lake Taihu under the optimal combinations of water transfer routes in the non-algal bloom seasons **a** and the algal bloom seasons **b**, respectively (Li et al. 2013b)

in central Anhui province, located in China (Fig. 4.6) (Huang et al. 2016). Lake Chao has been experienced an excess of nutrients loading during the past few years leading to worse water quality. Consequently, it resulted in severe algal blooms. The lake presents a spatial distribution of pollutants. In the western Lake area, the total phosphorus and nitrogen (0.66 and 7.63 mg/L) were higher than those in eastern Lake with a value of 0.34 and 5.29 mg/L, respectively (Huang et al. 2016). This presence of a relatively high level of nutrient in western part of Lake Chao was mainly due to the large nutrient discharge from Nanfei River with a large city (Hefei) in its watershed (Fig. 4.6).

4.3.1 Numerical Model

Xinanjiang and EFDC are the two numerical models which have been combined to get results. Water age and residence time were adapted to evaluate the pattern of water transport in different areas of Lake Chao. Water age has been defined previously. Residence time is defined as the time required for a material to remain in a defined region before being transported out of the defined region (Huang et al. 2016). Moreover, to define boundary of the residence time, in this study residence time is considered as the time needed for a material to be transported out of the square grid cell (1 × 1 km), rather than the lake (Fig. 4.7).

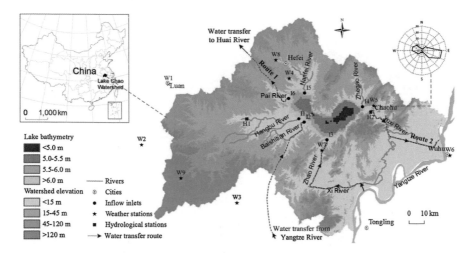

Fig. 4.6 Lake Chao watershed, Inflow Rivers, weather and hydrological stations (Huang et al. 2016)

Fig. 4.7 Water age and residence time in grid cells of a lake (Huang et al. 2016)

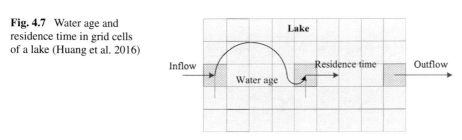

4.3.2 Results and Discussion

4.3.2.1 Changing the Transport Pattern Due to Water Transfer

The term "*hydrodynamic change region*" is used in this study to represent the area with significantly different hydrodynamic conditions between Sim_None and other simulations and therefore makes a comparison between them. The model configurations have taken into account four numerical scenarios to investigate the impact of the water transfer project on Lake Chao. These configurations are named Sim_None, Sim_Tran1, Sim_Tran2, Sim_Wind, and are presented in Table 4.2 with their specificities. According to Fig. 4.8, the water flow in the Western area was weaker than that in the Easter area, this because water outflowed through Yuxi River in the Eastern part. Water velocity around the western area is relatively large due to the presence of island adjoining it.

Configurations Sim_Tran1 and Sim_Tran2 highlighted the water transport change due to water transfer from the different routes. The two latter configurations brought about important changes in the vertically-averaged water velocities compared with

Table 4.2 Model configurations of four numerical simulations to investigate the impacts of the water transfer project (Huang et al. 2016)

Model configurations	Sim_None	Sim_Tran1	Sim_Tran2	Sim_Wind
Precipitation	0	0	0	0
Evaporation	0	0	0	0
Inflow and outflow discharge(m³/s)[a]	Hangbu River: 30.22 Baishitian River: 40.84 Zhao River: 7.82 Zhegao River: 7.96 Nanfei River: 24.67 Pai River: 7.99 Yuxi River: −119.50	Hangbu River: 30.22 Baishitian River: 56.69 Zhao River: 23.69 Zhegao River: 7.96 Nanfei River: 24.67 Pai River: 7.99 Yuxi River: −151.21	Hangbu River: 30.22 Baishitian River: 56.69 Zhao River: 23.69 Zhegao River: 7.96 Nanfei River: 24.67 Pai River: −23.72 Yuxi River: −119.50	Hangbu River: 30.22 Baishitian River: 40.84 Zhao River: 7.82 Zhegao River: 7.96 Nanfei River: 24.67 Pai River: 7.99 Yuxi River: −119.50
Wind condition	No Wind	No Wind	No Wind	2.5 m/s (eastern Wind)[b]

[a]Estimated by xinanjiang model and adjusted by the water transfer strategies
[b]Average wind conditins between 2011 and 2013

(a)   **(b)**   **(c)**

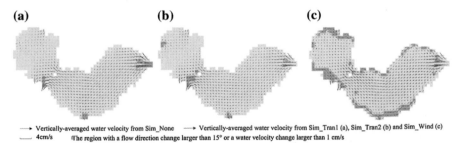

→ Vertically-averaged water velocity from Sim_None   → Vertically-averaged water velocity from Sim_Tran1 (a), Sim_Tran2 (b) and Sim_Wind (c)
—— 4cm/s   The region with a flow direction change larger than 15° or a water velocity change larger than 1 cm/s

Fig. 4.8 Comparisons of vertically-averaged water velocities from Sim None and other three simulations (Sim Tran1 (**a**), Sim Tran2 (**b**) and Sim Wind (**c**)) (Huang et al. 2016)

Sim None. Sim_Tran1, Sim_Tran2, and Sim_Wind had different hydrodynamic conditions compared to Sim_None (Fig. 4.8). The areas near the shore were the main place of hydrodynamic change region. Inflows and outflows were the principal causes of hydrodynamic change in configurations Sim_Tran1 and Sim_Tran2, as to the wind conditions, it derived from a high amount of circular water current in the near-shore areas (e.g., the southern Lake Chao in Fig. 4.8c) subsequently, the water velocity in the near-shore areas.

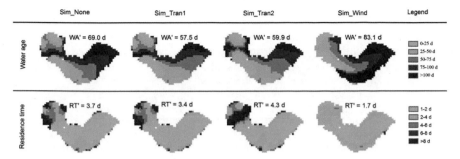

Fig. 4.9 Spatial water age and residence time in Lake Chao from Sim None, Sim Tran1, Sim Tran2 and Sim Wind. WA' and RT' represent the average water age and residence time in Lake Chao (Huang et al. 2016)

4.3.2.2 Water Age and Resident Time

Water transfer project has improved dilution of pollutants into the lake. Accordingly, both Sim_Tran1 and Sim_Tran2 were characterized by a smaller water age than Sim_None. Although resident time was different. Sim_Tran2 presented a resident time larger than Sim_None mainly due to the flow direction, which was from west to east. Shown in Fig. 4.6, using route one would hamper water in the Western area of Lake Chao to flow out through Yuxi River.

The effect of wind on the lake was significant, specifically easterly wind. It accelerated the water flow intensity, and therefore Sim Wind had a resident time shorter than that of Sim_None. However, Sim_Wind had water larger than Sim_None, implying that the easterly wind hampered the dissolved substance to be expelled out of Lake Chao. Although the mean hydraulic retention time of Lake Chao was 207 days, the computed average residence time (Fig. 4.9) was less than 5 days. Based on the definitions of water age and residence time (Fig. 4.7), the shorter water age and residence time are, the better the benefits for water exchange would be, including pollution dilution and removal by water flow in the lake.

4.4 Conclusion

Water transfer could thus be an emergency strategy to mitigate environmental disasters (e.g., serious algal blooms) in a particular area by accelerating the harmful substances to flow out of the lake. The different single water transfer routes had various effects on water age and resident time in different lake regions. It is vital to operate each water transfer route properly to set the optimal combination of the transferred routes and to improve the lake's water exchange capibility with a minimum economic cost and environmental impact.

References

Huang J, Yan R, Gao J, Zhang Z, Qi L (2016) Modeling the impacts of water transfer on water transport pattern in Lake Chao, China. Ecol Eng 95:271–279

Li Y, Tang C, Wang C, Anim Do YuZ, Acharya K (2013a) Improved Yangtze River Diversions: Are they helping to solve algal bloom problems in Lake Taihu, China? Ecol Eng 5:104–116

Li Y, Tang C, Wang C, Tian W, Pan B, Hua L, Lau J, Yu Z, Acharya K (2013b) Assessing and modeling impacts of different inter-basin water transfer routes on Lake Taihu and the Yangtze River, China. Ecol Eng 60:399–413

Zhuang W (2016) Eco-environmental impact of inter-basin water transfer projects: a review. Environ Sci Pollut Res 23:12867–12879

Chapter 5
Water Transfer to Achieve Environmental Issues: Waterfront Body

5.1 Introduction

Waterfront is a land or a dock area. It is located directly on a body of water such as a river, lake or ocean. It exists in many countries in the world. Concerning lake, the waterfront has a strong connection with the water body whereby it shares its water. The formation set by the waterfront lake and the external rivers constitutes a system in the environment. Several waterfront lakes exist in China, especially around the Yangtze River (Wang et al. 2014). Depending on their geographical location and the rainfall season, those waterfront lakes are relatively influenced by tides coming from Yangtze River. By carrying sediments and pollutants from Yangtze River to the waterfront lake, these tides are the drivers of some changes in the waterfront lake. For example, in flood and dry season, Waterfront Lake faces important changes in sediment concentration for the excessive water exchange and the decrease of environmental capacity for the insufficient water exchange. Besides, owing to an unbalance water exchange, Waterfront Lake suffers from some environmental problems with external rivers, such as sediment deposition, water quality deterioration, heavy metal pollution, and aquatic ecosystem destruction. It is, therefore, important to address water control between the waterfront lake and the external river. That is, its water quantity and quality should be assessed to solve these environmental problems. The quality is closely related to Heavy metals, suspended solids characterised by their high toxicity, persistence, hard degradation, and easy accumulation, etc.

A water regulation scheme is put forward for the Inner Lake, a typical waterfront lake located in the middle (lower) of the Yangtze River, north of Zhenjiang (Wang et al. 2014). The heavy metal Pb is selected as a typical factor for its high ecological risk through field measurements, laboratory experiments, mathematical simulation, and other methods. In this study, the variations of the concentration of *Pb* pollution load are quantitatively assessed in some years of reference prior to and post artificial water regulation. In the following parts, we successively present the *Pb* pollution into the inner lake, and the last part, the effect of water transfer on Waterfront Lake.

5.2 Study Area

Zhenjiang city which is located at the convergence of the Yangtze River and the Grand Channel is a typical waterfront city of the Changjiang Delta in Jiangsu Province, China. Its main waterfront body named Neijiang represents the Inner Lake. The latter is linked to the upstream of the Yangtze River through the leading channel (3000 m in length and 300 m in width) and flows to the downstream at the Jiaonan Gate. The inner Lake faces the impacts of tides from Yangtze River which appear twice a day (Wang et al. 2014). To avoid hydrodynamic changes and some environmental issues into the inner lake, two sluices gates were built Fig. 5.1.

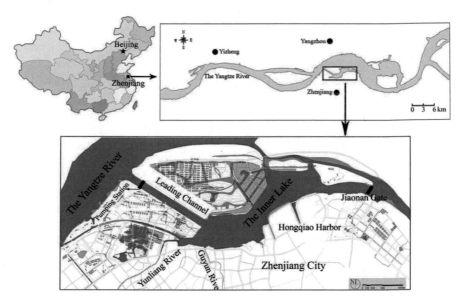

Fig. 5.1 General view of the study area (Wang et al. 2014)

5.3 The Inner Lake: *Pb* Pollution

5.3.1 The Pattern of Pb in the Sediment

Sediments present in the inner lake have a relative concentration of *Pb*. Figure 5.2a shows the different points where samples were taken. According to the method, samples processing used by Wang et al. (2014) associated to Lagrange's interpolation, the spatial distribution of *Pb* in the sediment of the Inner Lake is depicted in Fig. 5.2b. The presence of *Pb* was approximately 33.4% higher in the south of the lake than that in the north. This was caused by the location of Zhenjiang city (south of the lake) and subsequently by industrial and urban non-point source pollution. The average content of *Pb* in the whole lake was about 31.97 mkg^{-1}, which was above the background value of 26.03 mg kg^{-1} (Wang et al. 2014).

5.3.2 Mathematical Models for Migration and Transformation of Pb

The 2D equations which characterise the suspended solid and the *Pb* migration are:

$$\begin{cases} \frac{\partial h}{\partial t} + \frac{\partial (hu)}{\partial x} + \frac{\partial (hv)}{\partial y} = 0 \\ \frac{\partial hu}{\partial t} + \frac{\partial (hu^2 + gh^2/2)}{\partial x} + \frac{\partial (huv)}{\partial y} = gh(s_{0x} - s_{fx}) + hfv + hF_x \\ \frac{\partial hv}{\partial t} + \frac{\partial (huv)}{\partial x} + \frac{\partial (hv^2 + gh^2/2)}{\partial y} = gh(s_{0y} - s_{fy}) + hfu + hF_y \\ \frac{\partial S}{\partial t} + \frac{\partial (huS)}{\partial x} + \frac{\partial (hvS)}{\partial y} = \frac{\partial}{\partial x}\left(D_x h \frac{\partial S}{\partial x}\right) + \frac{\partial}{\partial y}\left(D_y h \frac{\partial S}{\partial y}\right) + E - D \\ \frac{\partial (hC_{AS})}{\partial t} + \frac{\partial (huC_{AS})}{\partial x} + \frac{\partial (hvC_{AS})}{\partial y} = \frac{\partial}{\partial x}\left(E_x h \frac{\partial C_{AS}}{\partial x}\right) + \frac{\partial}{\partial y}\left(E_y h \frac{\partial C_{AS}}{\partial y}\right) + \alpha w B(S - S_*) \bullet \Delta N \end{cases}$$

$$(5.1)$$

In these equations; *h* is water depth; *t* is time; u and *v* are the depth-averaged velocity components in the *x* and *y* directions respectively; *g* is acceleration due to gravity; s_{ox} and s_{fx} are the bed slope and friction slope in the *x*-direction; s_{oy} and s_{fy} are the bed slope and friction slope in the *y*-direction; F_x and F_y are the friction force components in the *x* and *y* directions, they can reflect the wind stress;

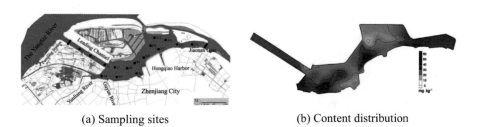

(a) Sampling sites   (b) Content distribution

Fig. 5.2 Investigated sites and content distribution of *Pb* in the Inner Lake (Wang et al. 2014)

f is Coriolis parameter. S is suspended solids concentration; Dx and D_y are the dispersion coefficient of suspended solids in the x and y directions under dynamic condition; E-D is the source-sink vector which reflects the rising and subsidence net flux of suspended solids. C_{Ax} is the concentration of Pb in water; E_x and E_y are the dispersion coefficient of Pb in the x and y directions under dynamic condition; S_* is the source-sink vector; B is the width of flow cross-section; ΔN is the value of sedimentation and resuspension process on Pb concentration which is based on laboratory experiments (see Wang et al. 2014). Gambit software and the established mathematical model were used to analyse Pb pollution load fluctuation before and after artificial water regulation (Wang et al. 2014). Moreover, three states which represent the hydrological condition of the Yangtze River have been taken as a basis, namely, the high water year (1998), the common water year (2001), and the lower water year (2004). Using the numerical simulation of the water exchange processes under specific conditions the exchanged water has been quantified.

5.3.3   Water Operation Schemes of the Inner Lakes

In Fig. 5.3 the numbers represent different operation schemes and their means are as follows "1" represents "*Self-introducing*". This mode is a condition in which water transfer from the Yangtze River is done naturally, owing to the high level of the latter River. "2" represents "*Pumping*". This mode is generally used with the help of water pumps which force water exchange. This occurs when the level of the Yangtze River is relatively low. "0" represents "Keeping static condition". This mode is always required after water exchange for the sedimentation and the increase of water transparency. The days for self-introducing in the high-water year are fewer than that in the other two years because the water level of the Yangtze River in a high-water year is higher and the one-time exchanged water quantity is more (Wang et al. 2014).

5.3.4   Variation of the Concentration of Pb Pollution Load

The variations of the concentration of Pb pollution termed here by its fluctuations where two conditions, namely, natural and artificial have been distinguished, is shown in Fig. 5.4. In natural condition, the average Pb pollution loads in the three relative water years, high, common, and low were respectively 33.32×10^4, 13.19×10^4, and 14.59×10^4 g/d. Pb pollution load was higher in the high water year than that of the others, representing about twice of that of them. The highest daily pollution loads in three typical years were respectively 113.88×10^4, 59.97×10^4 and 68.10×10^4 g/d. There are similarities of Pb pollution load fluctuations within different typical years, as the pollution load is higher in flood season (June to October) than that in the dry season (November to May). In the dry season, the average Pb pollution loads were

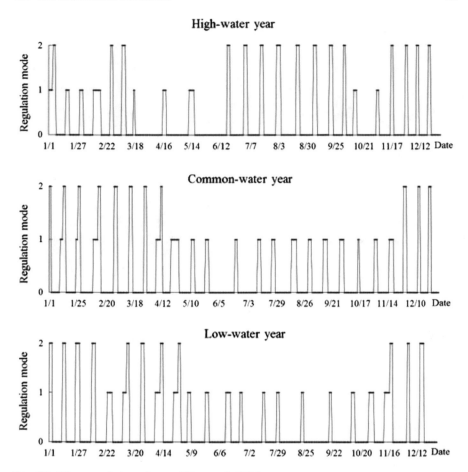

Fig. 5.3 Water regulation schemes (Wang et al. 2014)

8.08×10^4 g/d in the high-water year, 4.59×10^4 g/d in the common-water year, and 5.18×10^4 g/d in the low-water year. Whereas in flood season it was 80.12×10^4 g/d in the high-water year, 29.73×10^4 g/d in the common-water year, and 35.87×10^4 g/d in the low-water year. During the dry season the *Pb* pollution loads of the three relative water years presented similar patterns, meanwhile, in flood season, it was higher in high water year than that of the others. In the artificial condition, the *Pb* pollution load in the three years of reference, high, common, and low amounted respectively to 11.63×10^4, 6.00×10^4, and 5.42×10^4 g/d. Comparing to the natural condition, *pb* pollution loads decreased by about 65.12, 54.53, and 62.89%. The decrease is more significant in flood season, while it is less in the dry season. The results of *Pb* pollution load fluctuations in different typical years before and post artificial water regulation are shown as in Fig. 5.4.

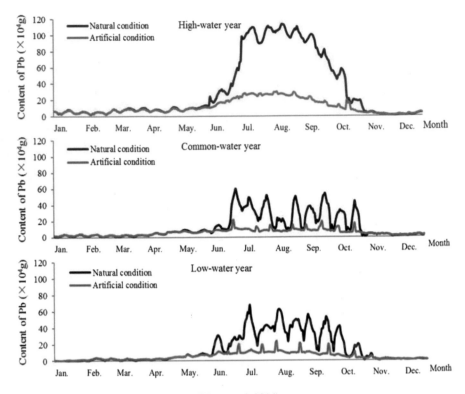

Fig. 5.4 *Pb* pollution load fluctuations (Wang et al. 2014)

5.4   Effects of Water Transfer in the Waterfront Body

5.4.1   Model Establishment

There is an uneven distribution of water transfer within the inner Lake owing, for example, to water season (high-low). This leads to a high level of complexity of the transformation mechanisms of water quality, suspended sediment, and water transparency. The use of a 2-D unsteady numerical model with different operation schemes of these environmental factors may help to identify water exchange. The model is an important basis to address a rational water operation scheme to achieve the anticipated environmental objects. The mathematical models describing water quantity and water quality, Suspended Sediments Transportation, algae Growth, and Submerged aquatic plant restoration are from Wang and Pang (2008).

5.4.1.1 Water Transparency Model

Water transparency which depends on many factors such as solar radiation, optics attenuation coefficient, physic-chemical characteristics, composition and concentration of suspended solids, and meteorological conditions etc. (Wang and Pang 2008), is an important parameter describing water optical status and assessing water eutrophication. According to the long term field investigation from Apr. 2004 to Oct. 2006 in Neijiang, it was found that suspended sediment concentration is the main factor influencing water transparency.

5.4.2 Mathematical Equations

The different equations of water flow, water quality, suspended sediment transportation, and algae growth were combined to be calculated. Hence, they can be written as the following unified form:

$$\frac{\partial q}{\partial t} + \frac{\partial f}{\partial x} + \frac{\partial g}{\partial y} = b(q) \tag{5.2}$$

where q is the vector of the conserved physical quantities; $f(q)$ and $g(q)$ are respectively the flux vectors in the x and y directions; $b(q)$ is the source-sink vector; the detailed expressions are as follows:

$$
\begin{aligned}
q &= (h, hu, hv, hC, hS, hN, hP, h(C_{chl-a})^T \\
f(q) &= (hu, hu^2 + gh^2/2, huv, huC, huS, huN, huP, huC_{chl-a})^T \\
g(q) &= (hv, huv, hv^2 + gh^2/2, hvC, hvS, hvN, hvP, hvC_{chl-a})^T \\
b(q) &= (0, gh(s_{ox} - s_{fx}), gh(s_{oy} - s_{fy}), \nabla \bullet (D_i \nabla(hC)) - khC + \\
&\quad S_c, \nabla \bullet (D_i \nabla(hC)) + E - D, \nabla \bullet (D_i \nabla(hN)) + S_N, \nabla \bullet (D_i \nabla(hP)) \\
&\quad + S_p, \nabla \bullet (D_i \nabla(hC_{chl-a})) + S_{chl-a})^T
\end{aligned}
\tag{5.3}
$$

superscript T is the transposing operator; ∇ is the gradient operator.

It is required to undertake a field work investigation in order to verify and validate a particular numerical model. For this study, the field works on Neijiang were done over two typical tidal cycles in the flood season (Aug. 18, 2004) and the dry season (Apr. 21, 2004). The boundary and initial conditions and the selection of parameters such as turbulent viscosity coefficient, sediment bulk density, dry-sediment bulk density, dispersion coefficient, algae death rate, and algae respiration rate etc. are from Wang and Pang (2008).

The calculated values and collected data were quite correlated with relative errors ranging from 10 to 15%. Accordingly, the established model can precisely reflect the migration and transformation laws of the environmental factors under different dynamic conditions in Neijiang.

5.5   Investigation on Water Quantity Operation

5.5.1   Water Quantity Operation

The principal purpose of the water transfer from the Yangtze River to Neijiang is to balance water exchange, improve water quality, control suspended sediment deposition, increase water transparency, and partially restore submerged aquatic plant (Wang and Pang 2008). Therefore, it is important to properly operate water transfer in Neijiang. The basic water operation schemes consisting in different points are as follows: (1) during the flood season, the suspended sediment concentration of the Yangtze River is high, the inflowing water should be closely controlled to monitor sediment deposition and enhance water transparency. (2) In dry season water level is low, and it is insufficient to maintain water exchange, so the pumping station located at the leading channel should be used to reinforce water exchange, to improve water carrying capacity and meet environmental flow requirement. (3) As every natural plant, the submerged aquatic plant presents in Neijiang has its natural water requirements to grow. Water quantity, therefore, should be a specific parameter to consider in different growth period of submerged aquatic plant, in return water level of Neijiang. (4) After operating, the modes of water exchange in Neijiang include three types: Self-introducing, Pumping, Keeping static condition (described previously). (5) Lastly, during flood seasons, the water operation schemes must be in a similar line of the flood control plan of Zhenjiang City.

5.5.2   Pattern of Water Quantity

Water transparency is characterised by the level of sediments concentration into the water body. To observe an increasing water transparency, the mode "*keeping static condition*" should be observed in order to foster sediments deposition. Evidently, the time required to reach a relatively good level of water transparency is driven proportionally towards the same direction by the initial level of sediment concentration. This indicates the improvement of the pollution discharge correlated to the gradual rise of the resting time and the degradation of the water quality in major water body of the City Rivers in Zhenjiang. To reasonably increase water transparency and improve water quality at the same time, the following condition should be met:

$$C_T \leq C_S \qquad (5.4)$$

Where C_S is the water quality concentration after the resting time T; C is water quality standard (the III grade of GB3838-2002 of China).

5.6 Environmental Effects Forecast After Water Quantity Operation

The use of the environmental and ecological numerical model, the transformation processes of water quantity, suspended sediment, and water quality and water transparency are simulated, then the environmental effects estimated. Note that these are done post water operation schemes for particular years.

5.6.1 Estimation of Water Quantity and Suspended Sediment

According to the calculated results, the effect of water transfer in Neijiang results in a decreasing amount of suspended sediments in the different level years. In the high water year basis, water operation scheme has a significant impact on the exchanged water quantity and sediment deposition. The last two decreased by 96.7 and 86.3% than that under the present condition, respectively. From a comparative perspective, the high-water year has a lower sediment concentration than that in both common and low-water year (7.41×10^4 and 6.43×10^4, respectively, expressed in $m^3 \ a^{-1}$). Figure 5.5 gives the results in the three different years.

	Under present condition	After water operation
High-water year	15.46	0.51
Common-water year	4.534	0.443
Low-water year	5.346	0.47

	Actuality	Water operation
High-water year	802	109.8
Common-water year	530	74.1
Low-water year	479.3	64.3

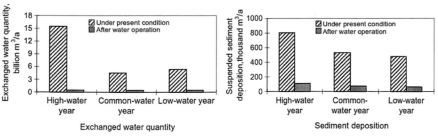

Fig. 5.5 Comparison of water exchange and sediment deposition before and after water operation in Neijiang (Wang and Pang 2008)

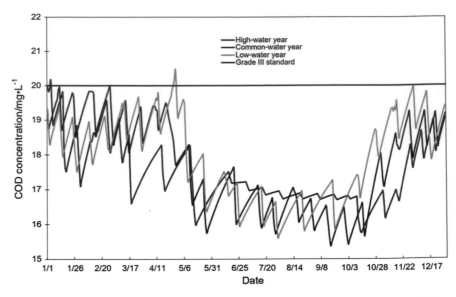

Fig. 5.6 COD processes after water operation schemes in typical years (Wang and Pang 2008)

5.6.2   *Water Quality Estimation*

As previously mentioned water transfer from Yangtze River into Neijiang can be helpful. The results showed that after water exchange is finished and the quiescence in Neijiang is done, the pollutant concentration decreases gradually. However, the sediment concentration is checked prior and post water operation schemes before executing the next time of water exchange. Hence, water quality in Neijiang was improved and can meet the expected standard. Figure 5.6 shows as an example of the COD processes post water operation schemes in particular years.

5.6.3   *Assessing Water Transparency*

The ranging averaged water transparency in Neijiang is 10 cm to 15 cm. When operating the water exchange increases the suspended sediment in Neijiang. However after the operation scheme, the flux of sediments is lowered and, consequently, its water transparency is enhanced significantly. The results give the average water transparency in the three water level at a year basis. In the high-water, common-water, and low-water the water transparency is respectively 69.3, 70.8, and 69.8 cm. These previous values can reach the anticipated goal, 50 cm. We note that the initial period of water exchange is marked by the decrease in water transparency and, then,

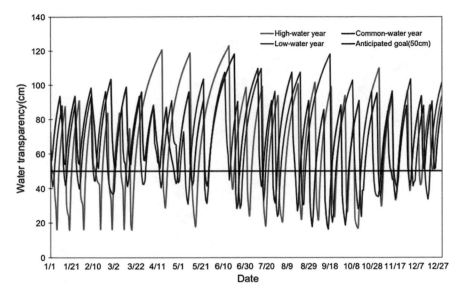

Fig. 5.7 Water transparency processes after water operation schemes in typical years (Wang and Pang 2008)

it becomes better after a relatively long resting time. Figure 5.7 shows the variations of the water transparency post water operation.

5.6.4   Assessing Submerged Aquatic Plant Restoration

As a submerged aquatic plant, Vallisneria spiralis is taken here as an example to show the effect of water transfer scheme in Neijiang. The results showed that the water exchange creates a favourable living conditions for plants and, therefore, Vallisneria spiralis may benefit from it, especially in the north-eastern bottomland and the south part of Neijiang. The entire areas of these regions in high, common, and low water year are 0.67, 1.06, and 1.01 km^2. The spatial distribution of Vallisneria spiralis restoration area is shown in Fig. 5.8.

5.7   Conclusion

Waterfront Lake is more often linked to environmental issues due to the external rivers and sometimes the land just nearby Waterfront Lake (waterfront land). In China, the inner lake close to Zhenjiang city also faces series of environmental problems, such as unbalance water quantity exchange, water quality deterioration, suspended sediment

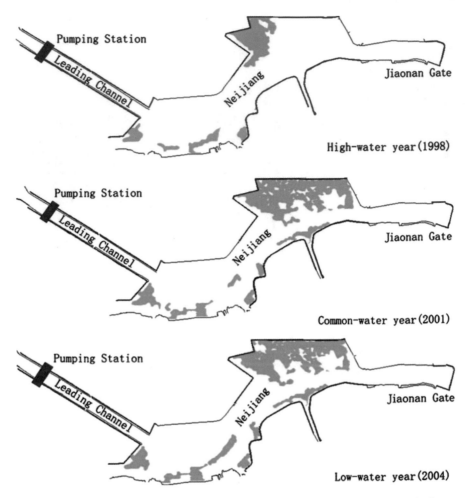

Fig. 5.8 Distribution of Vallisneria spiralis restoration area after water operation in typical years (Wang and Pang 2008)

deposition, water transparency depravation, and frangibility of the ecological system, etc. In particular conditions, water transfer into the inner lake has presented some significant advantages.

References

Wang H, Pang Y (2008) Water quantity operation to achieve multi-environmental goals for a water-front body. Water Resour Manag 23:1951–1968
Wang H, Zhou Y, Pang Y, Wang X (2014) Influence of water regulation on internal heavy metal load for a tide-influenced waterfront lake. Adv Mater Res 23(10):1951–1968

Chapter 6
Case of Physical Water Transfer from Yangtze River: Different Routes

6.1 Introduction

We all know that water is essential for life, but its access is still much more difficult depending on our geographic positions. Some areas are relatively dry, and there is about 80% of humans who live in a state of water insecurity (Wilson et al. 2017). Although at a global scale, water is abundant inferring that the issues of water scarcity are due to spatial and temporal uneven distribution of water and the mismatch between supply and demand. Many solutions exist to tackle water scarcity issue. One of them is to move water from water abundant place to water-scarce place by using water channels. China is a classic example of a country where a gap between water supply and demand exists. The country is water abundant in the south and water scarce in the north, so through SWTP, water has been transferred from the Yangtze River to the North. The SNWTP includes four of the seven major rivers in China and affects almost one-third of China's land mass (Zhang 2009) and consists in three routes (east, central and west route). Because of their different points of water supply and the location of their channel, (and other factors) the transferred water through the routes would have different features both in quantity and quality. The SNWTP, therefore, shares not only water but also some problems both in the receiving and in the supplying area. In the following sections, the different routes are presented separately, followed by their impacts.

Y. Li et al., *Addressing the Uneven Distribution of Water Quantity and Quality Endowment*, SpringerBriefs in Water Science and Technology, https://doi.org/10.1007/978-981-13-9163-7_6

6.2   Description of the Different Routes

6.2.1   The Eastern Route

The eastern route commences at Sanjiangying in Jiangsu Province and diverts water directly from the main course of the lower Yangtze River. It is the most technically feasible, since the channel uses the existing Grand Canal waterway. The latter way was originally used as a trade artery connecting Hangzhou to Beijing. The transferred water supplies the Northern provinces, including Jiangsu, Shandong, Hebei, Tianjin, and areas in the Huai River basin located between the Bengbu Water Gate and the banks of the Xinbian River in Anhui Province (Rogers et al. 2016). The route passes through four big lakes: the Hongze, Luoma, Nansi and Dongping. The total length of the main canal is about 1150 km, and the part in the south of the Yellow River is about 660 km long (He et al. 2010). The designed capacity (to 2020) of the transfer from the Yangtze River and the inflow to Hongze Lake is 1000 m^3/s. Thus, the northward flow would decrease gradually to about 180 m^3/s at Tianjin. The capacity of the reservoir along the transfer route is approximately 753108 m^3 in the south of the Yellow River and 123108 m^3 in the north of the Yellow River, for a total capacity of 873108 m^3. Note that some parts of the canal, particularly, near Shanghai and Hangzhou, are heavily used by barge traffic. One important challenge of this portion of the water transfer project, is the gravity. The Grand Canal's channel bed elevation above sea level gradually increases over the first (southern) two-thirds of the 1156-km length (Magee 2011), rising roughly about 130 ft (40 m) from Hangzhou (the southern terminus of the canal) to Jining in Shandong Province, where it reaches its maximum elevation. Water pump is, therefore, required to guarantee the flow of Yangtze River water northward (uphill). The use of water pumps requires energy, and eastern China has suffered from acute electric power shortages in recent years (Magee 2011). A total of almost 70 pumps are included in this route.

6.2.2   The Middle Route

It transfers water from the Danjiangkou reservoir on the Han River, which is a large tributary to the middle reaches of the Yangtze River, to Hubei, Henan, and Hebei provinces, and ultimately to Beijing and Tianjin, as well as to the western part of the North China Plain (He et al. 2010). According to the planned project, the total length of the main canal is about 1246 km, of which about 482 km located in the south of the Yellow River and about 764 km located in the north of the Yellow River (Magee 2011) (Fig. 6.1).

If we can make a slight comparison between the eastern and the middle route, the latter route of the SNWT is the more attractive owing to the fact that water would flow naturally in most areas of the channel. In addition, there is no existing channel for the central route, so the potential for social, economic, and ecological disruption

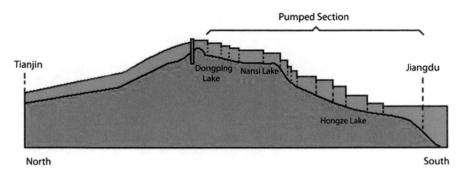

Fig. 6.1   Elevation profile of the Eastern route (Magee 2011)

resulting from constructing the channel is far greater than that of the eastern route. Supplemental water may be pumped up to Danjiangkou from the reservoir created by the Three Gorges (Sanxia) dam or from downstream Shashi. An alternative scheme is to pump water diverted from the Hanjiang River at Xiangfan into the Tangbai River and then along the Bai River into the main canal (Yang et al. 2015). This transfer scheme is mainly designed to address municipal and industrial water uses in the cities of Beijing and Tianjin, and provinces of Hebei, Henan, and Hubei. This includes supplementing water supplies for municipal, industrial, and agricultural use, as well as for ecological demands. This channel would follow the route of the Beijing-Guangdong railway line, traversing the Yue, Huai, Yellow, and Hai River watersheds along the way. The primary trunk from Danjiangkou to Beijing will require the construction of almost 700 new road bridges, with the demolition of some 1660 pre-existing structures along the channel route.

6.2.3   The Western Route

The western route is formed by many disjoint segments. Among all the three routes, it presents a technical difficulty which resides in the controversy of its feasibility. This is due to in the first place, the route would traverse ecologically and culturally diverse areas of western China, potentially posing unwarranted risks in terms of species loss and challenges to human livelihoods, sites of cultural importance, and traditions. In the second place, the problem of altitude rises. This would imply to direct water flow uphill at points over the eastern Himalayas, requiring largescale enigeneering projects, as well as pumping and syphoning to lift water over ridges, all of which would take place in rough terrain (Magee 2011). Like for the eastern route water pumps would require electricity in the southwestern part of the country. Lastly, in the long run, there is less insight into the understanding of the impacts of major inter-basin transfers (Magee 2011). This is particularly meaningful in the upper reaches of the Yangtze and its tributaries, where transferred volumes would comprise a greater percentage of in-stream flows in the Yangtze itself.

The western route would divert water from the upper reaches of the Yangtze River to the upper reaches of the Yellow River to provide water for northwestern China. The investigation about the possibilities of realising western routes has covered an area of more than 600000 km^2 in Qinghai, Gansu, Sichuan, and Yunnan provinces (Magee 2011). Several schemes for channelling water have been investigated. The magnitude of the western route scheme would be very large and intricate, requiring the construction of 50000 km of canals diverting 500 billion cube meters of water from the big southwestern river basins. The principal objective of the western route would be to supplement water in the Yellow River and its upper tributaries, primarily to meet industrial, municipal, and agricultural water demands in Qinghai, Gansu, Ningxia, Inner Mongolia, Shaanxi, and Shanxi Provinces.

6.3   Impacts of the Different Routes of the SNWTP

6.3.1   Eastern Route

Water transferred through the Eastern Route would flow from the lower reach of the Yangtze River will pass through and impound four lakes (Hongze, Luoma, Nansi, and Dongping). The water level in the lakes will increase at rates of 0.5 and 2–3 m in the Hongze and Dongping Lakes, respectively (Zhang 2009). Consequently, the reversed hydrologic regime may occur. The water level of these lakes is normally lower in winter and higher in summer, but this state will be changed after implementing the SNWTP project.

Ecological impacts of Lake impoundment, water quality degradation along the canal, secondary salinisation in the receiving areas, and invasion of alien species are the principal issues concerning the eastern route (Zhang 2009). It has been found out that, there is a possibility of the development of parasitic diseases such as schistosomiasis owing to the share of water between the northern areas and the infected area in Jiangsu Province. For example, over the period 1989–1998, a total of 7772 cases of acute schistosomiasis infection were reported in Hubei Province alone. We also noted that drifting pieces of reed carrying snails from the infected area could lead to the development of new snail habitats in the riparian area of lower reaches of the Yangtze River (Shao et al. 2003). In addition, the coastal areas may face the seawater intrusion which leads to the salinisation of soils in the water receiving areas (He et al. 2010). This route covers on one of the most developed regions in China. One issue related to it is the water quality along the channel. The population growth and the increasing demands of water for industries have led to wastewater discharge, agricultural pollution, and municipal wastes into the channel in return affecting its water quality (Zhang 2009).

6.3.2 Middle Route

There are several environmental issues associated to this project including soil salinisation caused by the rising groundwater table affecting water quality, slope instability of swelling clay and rock, seepage through the banks of the canal, and frozen heave problems (He et al. 2010). The central Route supplies the same water receiving areas (North Plain China) as the East Route does, so, its main environmental concerns also include secondary salinisation in the receiving areas and invasion of alien species (Zhang 2009).

The Middle Route passes through hundreds of rivers, canals, and streams and links relatively populous areas (Zhang 2009). One of them, the middle and lower Hanjiang River is an important economic corridor of Hubei Province. The average annual quantity of water resources in this region is about 19.4 billion m^3, accounting for 33.3% of the whole Han River Basin (Gu et al. 2012). The only huge reservoir is the Danjiangkou reservoir, and both precipitations and stormwater are uneven. Thus, the principal water source for the economic development of the middle and lower Hanjiang River comes from it. After the implementation of the water transfer route, part of the water will flow to the north of China, and the quantity and process of water from Danjiankou reservoir will be decreased and affected significantly. This will likely lead to the decrease of the capacity of water supply to the middle and lower Hanjiang River. Moreover, many factors such as population growth, food demand, and industrial and municipal development are likely to induce water shortages and consequently water stress in the supplying area. Demand in this area may increase from 710 to 960 Mm^3 until 2030 (Gu et al. 2012). Besides, the annual average water level at the downstream of the Danjiangkou reservoir would drop by 0.52–1.31 m and 0.63–1.28 m during April–October and July–September respectively (Zhang 2009).

6.3.3 Western Route

The Western Route is far away from the East and central Routes, and it is in the least developed region of China. The primary environmental concerns surrounding construction of this route include geological disasters (e.g., earthquakes and landslides), disease propagation, and impacts on the riverine ecosystems of the upper Yellow River. Environmentally, how the transfer scheme will impact the Qinghai–Tibet plateau tundra is a little complicated to assess (He et al. 2010). The Western Route is planned to divert about 17 billion m^3 of water, which corresponds to less than 15% of the annual discharge of Tongtian, Yalu, and Dadu Rivers, all tributaries of the upper Yangtze River (Zhang 2009). Although, this amount represents 65% of the annual discharge at the transferring point, the planned water storage which will be used, could increase regional evaporation and in return precipitation by 0.4–0.7%. Thus, to some extent adds water availability. In the future, water-intensive sectors such as agriculture, industries, and domestic will increase their water demands by 40% from the current 56.2 billion m³/year to 78.4 billion m³/year. However, the surplus from the western route should be of a great contribution.

6.4   Assessing Environmental Impacts of Water Consumption in China

6.4.1   Method

The Characterization factor of impact for a given region and scenario have been used here according to the impact assessment method given by Lin et al. (2012). The method includes indicators such as generic "*midpoint*" to address water stress as well as "*endpoint*" factors based on the framework of the Eco-indicator 99(EI99) method. This impact factor can use water stress index (WSI) as the "*midpoint impact*", ecosystem quality (EQ), human health (HH), and resources (RS) as the endpoint impacts. It can also use an aggregated impact (AI) as an aggregated index. The previous index is defined as follows (Lin et al. 2012). EQ addresses the potential ecosystem damage and considers the limitations of net primary production due to water availability and precipitation amounts. RS is the water used above the renewability rate and estimates the potential energy required to replace the depleted water by a backup technology. HH is the potential impacts on human health due to a lack of water for agriculture and subsequent malnutrition-related health problems. The EI99 point is defined in Eco-indicator for the "hierarchism" perspective, and average weighting is selected as the unit for endpoint impacts. For instance, 1,000 EI99 points equal 1 person-year equivalent (in terms of impacts caused by the activities of an average European for one year). It should be noted that some indexes used are presented here, for the others describing the whole method see Lin et al. (2012).

6.4.2   Results and Discussion

Two scenarios have been taken into account. The "*before-project*" and "*after-project*" scenarios. The first one is considered in the year 2000, while the second represents the situation where the first stage of the SNWTP is completed, that is, 2014. The results considered only the eastern and middle route of the SNWTP due to the highly populated and developed regions that they covered.

6.4.2.1   Water Availability

The hydrological availability and withdrawal water for the two scenarios are shown in Fig. 6.2. The water availability in the North has increased in the after scenario of

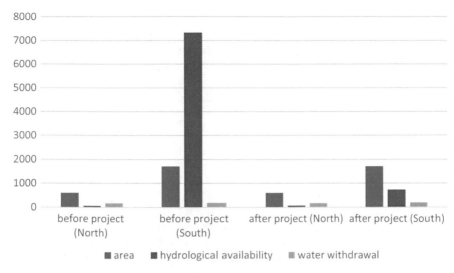

Fig. 6.2 Scenarios (adapted from Lin et al. 2012) *Note* Area (billion m^2), hydrological availability and water withdrawal (billion m^3)

about 7.38 billion m^3/yr of water transferred at the same time nearly the same amount is consumed in the southern regions. Obviously, water withdrawal in the South has increased.

6.4.2.2 Project Environmental Impacts of Water Consumption Embodied in Final Demands

Before-Project

Using the provincial water withdrawal, water consumption embodied in final demand has been assessed as well as its environmental impact for the two scenarios. Different sectors are obtained. In the before-project scenario, the total water consumption embodied in the final demands of the North is 33.1 billion m^3/yr., final demand in the South is 110 billion m^3/yr. both previous values present a relatively high difference. In the perspective of the environmental impact, the aggregated index of the North is 7.22 billion pts/yr., while that of the South is 10.7 billion pts/yr., for a ratio of 1.48 (Lin et al. 2012). According to that result, one unit of water consumed in the North causes more environmental impacts than the same consumption in the South.

The environmental impact embodied in the final demand of the North is not only due to the internal demand, because the amount of water consumption in one region is exported to the other as water is embedded in commodities. For instance, among the 7.22 billion pts, 126 million pts is caused by export to the South. Meanwhile, among the 10.7 billion pts of environmental impact embodied in the final demand of the South, 751 million pts is due to export to the North. Some sectors have the largest water-related environmental impact, namely the agriculture, livestock, forestry, and fishery sector. The former sector has a ratio of water consumption embodied in the final demand of about 41.9% in the North, and 49.0% in the South, due to its lower water consumption coefficient in the North than that in the South.

After-Project

The environmental impacts for the two scenarios (after-project scenario and before-project scenario) have been assessed. We noted that the characterisation factors are replaced by those of the after-project scenario (see Fig. 6.3 HH for instance). The decrease in water consumption coefficients of the North caused by the use of southern water. Comparing the two scenarios, the results showed that the impacts of HH and RS in the North face high decreases due to the smaller characterisation factors and decreased water consumption coefficients. While, the net decrease of the impact of EQ is less than HH and RS, because the characterisation factor of EQ remains unchanged. In return, the impacts of HH and RS in the South increase due to the increasing characterisation factors and the additional extraction of water, while the impacts of EQ in the South increase, because of the increasing extraction and transfer to the north, with the characterisation factor of EQ unchanged. In the perspective of the aggregated impact, the environmental impact decreases from 7.22 billion pts/yr. to 5.53 billion pts/yr. in the North (about 23.4%).

From the EI99 indicator, the SNWTP would save 2 million person-equivalent impacts in the North. This number, in terms of person-equivalents, is 0.94% of total impacts, as the total population of the North in 2000 was 180 million. An increase of 6.22% (from 10.7 to 11.4 billion pts/yr.) of aggregated impact is observed for the South. This increase (664 million pts/yr.), in terms of person-equivalents, is 0.13% of total impacts, as the total population of the South in 2000 was about 492 million (the population estimated here is only for the areas covered by the eastern and middle route of the SNWTP).

Overall, the environmental impacts induced by water consumption for the two scenarios (before and after) are significant. As such, the reduction of the environmental impacts of the South and North combined accounted for 5.74%. At last, the effect of the SNWTP in the perspective of environmental impacts induced by water consumption embodied in final demands is great.

6.5 Conclusion

Overall, the implementation of the SNWTP will alleviate water shortages in the northern part of China and provide much-needed water in northwest China for the rehabilitation of the degraded ecosystems. However, it might alter the ecosystems in

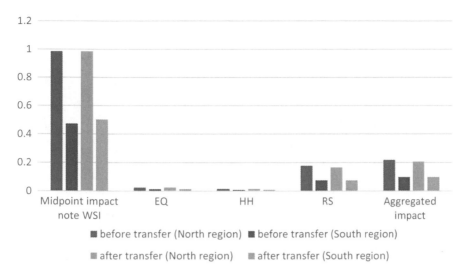

Fig. 6.3 Characterisation factors (adapted from Lin et al. 2012) *Note* WSI = water stress index; EQ = ecosystem quality; HH = human health; RS = resources; AI = aggregated impact; Pts/m^3 = points per cubic meter. (EQ, HH, RS, and AI are expressed in Pts/m^3 and except AI, they represent Endpoint impact)

the water supplying and receiving areas. Both water quantity and quality are the two most important features which should be included when assessing water transfer.

References

Gu W, Shao D, Jiang Y (2012) Risk evaluation of water shortage in source area of middle route project for south-to-north water transfer in China. Water Resour Manag 26:3479–3493

He C, He X, Fu L (2010) China's south-to-north water transfer project: Is it needed? Geogr Compass 4(9):1312–1323

Lin C, Suh S, Pfister S (2012) Does south-to-north water transfer reduce the environmental impact of water consumption in China? J Ind Ecol 16: 647–654. https://en.wikipedia.org/wiki/Journal_of_Industrial_Ecology

Magee D (2011) Moving the river? China's south–north water transfer project. In: Engineering Earth. Springer, Dordrecht, pp 1499–1514

Rogers S, Barnett J, Webber M, Finlayson B, Wang M (2016) Governmentality and the conduct of water: China's South-North Water Transfer Project. Trans Inst Br Geogr 41(4):429–441

Shao X, Wang H, Wang Z (2003) Interbasin transfer projects and their implications: A China case study. Int J River Basin Manag 1:5–14

Wilson MC, Li XY, Ma YJ, Smith AT, Wu J (2017) A Review of the Economic, Social, and Environmental Impacts of China's South-North Water Transfer Project: A Sustainability Perspective. Sustainability 9:1489

Yang Y, Yin L, Zhang Q (2015) Quantity versus quality in China's south-to-north water diversion project: a system dynamics analysis. Water 7:2142–2160

Zhang Q (2009) The south-to-north water transfer project of China: environmental implications and
    monitoring strategy. J Am Water Resour Assoc 45(5):1238–1247

Chapter 7
Virtual Water Transfer Within China: The Case of Shanghai

7.1 Introduction

Due to the spatial distribution of Water resources around the world, some regions/countries cannot provide for themselves. They, therefore, rely on either their water or supplying water from others, or both to meet their needs. The externalisation of water supply is carried out through trade of commodities and services, made between countries of interest. The direct or indirect use of water produces pollutant such as Carbon dioxide (CO_2) (Lin et al. 2014; Peters et al. 2011). For example, it is well known that most of the richest countries in the world, generate pollutants indirectly through trade in the exporting regions/countries (mostly in developing regions/countries) (Feng et al. 2014). This raises the fact that such pollution can affect both the producing and the consuming region/country.

Furthermore, consumption patterns can be of great impact on regional water stress level. Recently, it has been reported that trade can act as a mechanism whereby wealthy consumers shift local water quantity stress to the economically poorer exporters of goods and services (Zhao et al. 2016). China, being the second largest economy in the world, experiences different levels of water stress both in quantity and quality. Specifically, developed provinces within China have been in water scare conditions and have outsourced their water supply to meet their demands. Shanghai, the largest megacity in China, in a similar manner has externalised its water supply to relieve its water stress and to meet the growing needs.

This chapter focuses on Shanghai's shifting of water stress both quantity and quality among China's provinces. The multi-regional input-output model and the index water stress level have been used in this chapter.

7.2   Shanghai's Water Endowment

Shanghai is a megacity which covers an area of about 6,340.5 square kilometres, and located in the central coast in China. A large amount of its water supply comes from the Huangpu River which is a tributary of the Yangtze River. More specifically, until 2010 more than 70% of its freshwater supply came from the Taihu Lake via the Huangpu River. Though, Shanghai's water quality is relatively dependent on upstream flows. Recently, water quality of the upstream river (Taihu Lake) has been deteriorated due to the economic activities and the lack of monitoring in the upstream regions of Jiangsu and Zhejiang. According to China's water quality standard, grades above Grade III indicate poor water quality which is unsafe. Grade V indicates that the water is seriously polluted and not to be for any propitious use. In 2007 only 12.5% of the river met the surface water quality standard, whereas 56.7% of the length of the river was considered to be worse than the Grade V. The Huangpu River has become a channel to convey polluted water discharge from Shanghai and the upstream regions to the sea. Shanghai, facing this water severe pollution issues, three reservoirs have been built to make the Yangtze River as its principal water resource, the Qingcaosha, the Dongfeng xisha and Chenhang Reservoirs (Fig. 7.1).

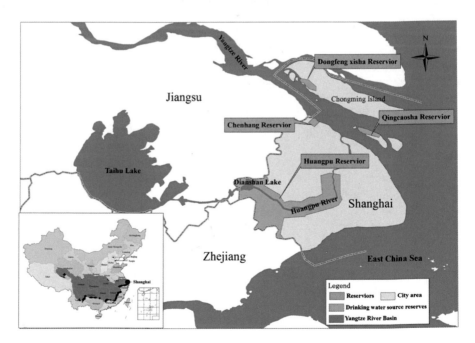

Fig. 7.1 Shanghai's water storage (Zhao et al. 2016)

7.3 Method and Data

The results are obtained from the multiregional input-output table developed by Feng et al. (2013) as a database. It is an aggregate of 30 industrial sectors within 30 provinces of mainland China. The method of calculating virtual water trade among regions and water embodied in trade have been taken from Zhao et al. (2016).

7.3.1 Water Stress Index

Water stress index is generally defined as the relationship between total water use and water availability. The closer water use is to water supply, the more likely stress will occur in natural and human systems. In this study, we distinguished water stress index in the perspective of quality and quantity. The water quantity stress index (I) is calculated as the ratio of water withdrawal (P, m^{3}yr.$^{-1}$) to annual renewable freshwater (T, m^{3}yr.$^{-1}$):

$$I = \frac{P}{T} \tag{7.1}$$

I represents the water stress index. Depending on the value it can take, and on the perspective of water, one region can be classified within four levels; Extreme ($1 < I$), Severe ($0.4 < I < 1$), Moderate ($0.2 < I < 0.4$), and No stress ($0.2 < I < 0.4$). The second perspective which concerns water quality stress index takes into account grey water footprint. The latter is to be understood as the amount of water required to assimilate pollutants load based on existing ambient water quality standard (Mekonnen and Hoekstra 2010).

$$I_q = \frac{G}{T} \tag{7.2}$$

I_q is the water quality stress index, and G is the grey water footprint. G is obtain from the following equation.

$$G = \max\left(\frac{L}{C_{max} - C_{nat}}\right) \tag{7.3}$$

Where L, C_{max}, C_{nat} are respectively the load of pollutants (ton yr.$^{-1}$), the ambient water quality standard (mg.l^{-1}), and the natural background concentration (mg.l^{-1}). From criteria suggested by Zeng et al. (2013), if I_q is less than 1, this implies T can assimilate the existing load of pollutants based on the local water quality standard. Hence, $I_q < 1$ characterises no stress. In contrast, if I_q is greater than 1, then freshwater availability is insufficient to dilute the polluted water. We then subdivide water stress into three classes according to the proximity of the results cluster, Extreme ($5 < I_q$), Severe ($2 < I_q < 5$), and Moderate ($1 < I_q < 2$).

7.4 Results and Discussion

As mentioned above, virtual water of different sectors has been highlighted. Sectors have been classified as a primary, secondary and tertiary industry. Agriculture is categorised as primary industry, while Coal Mining, Dressing and Construction are categorised as secondary industry, and sectors of Freight Transport, Warehousing, and Other Services are classified as tertiary or service industry. COD and NH_3–N have been taken as the main pollutants to evaluate the impact of Shanghai's consumption among provinces in the perspective of water quality induced water scarcity.

7.4.1 Shanghai's Consumption and Water Quantity

Shanghai was a net virtual water importer in different sectors in 2007. It has imported around 79% of its total water consumption in virtual form from other provinces. Some sub-sectors have significantly contributed to Shanghai's virtual water import, namely agriculture, food and tobacco processing, and hotel and catering. Agricultural products account for about 63.1% of its total virtual water import followed by food and tobacco processing accounting for about 15% (Fig. 7.2a).

The main flows of virtual water exported to Shanghai are shown in Fig. 7.3a. Xinjiang, Inner-Mongolia, Hebei, Anhui, Heilongjiang, and Jiangsu were the top exporting provinces, their net virtual water exported account for 4810 million m^3 which represents about 56% of Shanghai's virtual water import.

Note that the net flows larger than 500 million cubic meter and volumes larger than 35×10^3 tons are shown in Fig. 7.3a. From Fig. 7.3a, Shanghai has a water quantity stress above 1. This means that being at extreme level, it has overexploited its water resources to satisfy its needs. Besides, there are other provinces which are in the same situation as Shanghai. Most of the Northern provinces which are virtual water exporters are in severe water stress ($0.4 < I < 1$), while many in the South are in the moderate level. Furthermore, 13 provinces with extreme and severe water quantity stress are responsible for about 60% of net virtual water export to Shanghai. For instance, Xinjiang which is in severe water quantity stress exports 1629 million m^3 of virtual water to Shanghai, Hebei one of the water-scarce provinces participates in water trade by exporting 672 million m^3 of virtual water to Shanghai. It should be noted that those Northern provinces are the principal producers of agricultural products in China.

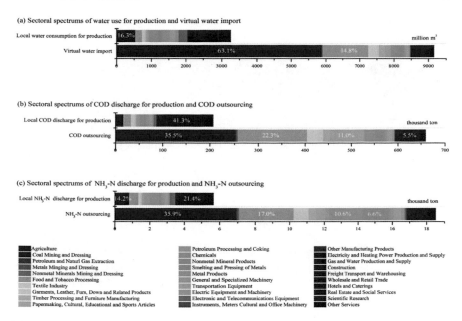

Fig. 7.2 Virtual water import and local water consumption of Shanghai in different sectors. Percentages indicate the shares of the sectors in total water consumption (Zhao et al. 2016)

7.4.2   Shanghai's Consumption and Water Quality

From Fig. 7.2b, c shanghai had in 2007, direct and indirect production of COD and NH$_3$–N of about 796,000 and 24,000 tons respectively. This production is mainly from the discharged wastewater of other provinces owing to Shanghai's consumption of commodities and services. The indirect pollution generated by Shanghai in other provinces amounted to 665, 000 tons of COD and 18,500 tons of NH$_3$–N accounting for about 83.5 and 77% of the total of COD and NH$_3$–N, respectively. Shanghai generated itself about 207,000 and 5,800 tons of COD and NH$_3$-N, respectively, for producing its goods and services.

The largest share of pollutants from external sources in 2007 was from agriculture, which amounted to 240,000 and 6,800 tons of COD and NH$_3$–H respectively, followed by metal mining and dressing, nonmetal minerals mining and dressing, food and tobacco processing. The last three aforementioned account for 22.3% of the total COD and 17% of the total NH$_3$–H among all sectors. Although agriculture was not the primary generator of pollutants in Shanghai, it was dominated by other sectors including hotels and caterings, freight transport and warehousing, and whole and retail trade. Many provinces located in northern China have relatively a water stress quality more than one (Fig. 7.3b). This means that, for the latter, in spite of being in water quality and quantity stress conditions, Shanghai has generated pollutants from them and continue to exacerbate the use of their water resources.

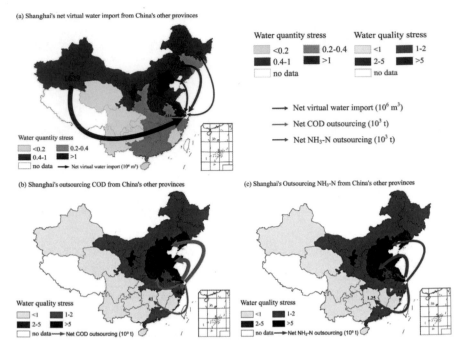

Fig. 7.3 a Shanghai's net virtual water import from other Provinces. The colours of the provinces indicate their water quantity stress status. The flows with arrows show the top net virtual water exporters to Shanghai. **b** COD in other provinces due to Shanghai's consumption (Zhao et al. 2016)

Shanghai has externalised its water source by importing goods and services and, therefore, has generated pollutants to the other provinces. Nineteen provinces in water quality stress accounts for 79% of net COD and 75.5% of net NH_3–N outsourcing from Shanghai. Shandong, Hebei, Zhejiang, and Henan endured Shanghai's net indirect generation of COD, while Anhui, Henan, Hebei, Zhejiang, and Jiangsu were the principal provinces where Shanghai's net NH_3–H is generated. Hebei is the most affected by virtual water import and pollutant outsourcing from Shanghai, and it is suffering water stress in terms of both quality and quantity.

7.4.3   Magnitude of Water Trade

Water intensity reflects the value of water through trade of commodities and services. In 2007, Shanghai to relieve its water stress level, exports low water-intensive goods and services to other provinces and imports water-intensive products from them (Zhao et al. 2016). Imported water intensity was 9 times higher than its exported water intensity, that is to say, 1 m^3 of water used in Shanghai on average can make 1000 CNY of goods and services exported to other provinces, which in turn can

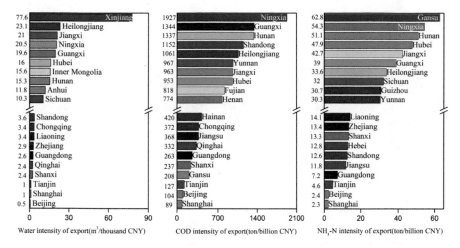

Fig. 7.4 water intensity of the main largest and lowest provinces in China (Zhao et al. 2016)

only produce 111 CNY of goods and services imported from other provinces. After Beijing, Shanghai had the least water intensity of export. The largest virtual water of export is Xinjiang province (Fig. 7.4).

Similarly, water intensity has been expressed in terms of water quality (Fig. 7.4). The water pollutant intensity is the direct and indirect water pollutant discharge per unit of trade (Zhao et al. 2016). In this case COD and NH_3-N intensity of imports (649 ton/billion CNY and 17.8 ton/billion CNY) were more than 7 times larger than the COD and NH_3-N intensity of exports (89 ton/billion CNY and 2.3 ton/billion CNY). Shanghai had both the lowest COD and NH_3-N intensity of exports among all Provinces. Ningxia had the largest COD intensity of exports (1927.3 ton/billion CNY), which was about 21 times larger than that of Shanghai. Gansu had the largest NH_3-N intensity of exports (62.3 ton/billion CNY), about 27 times larger than that of Shanghai.

7.5 Conclusion

Water endowment is relatively different among regions due to their geographic position. However, despite their poor water resources, some regions/countries have found ways to alleviate water stress-induced water scarcity within their boundaries. That is to say, virtual water. This latter has been directly or indirectly the generator of some issues both in the production and consumption side. The results obtained with Shanghai megacity showed that virtual water is indeed a way to relieve water for needs, but it is also to some extent the cause of accelerating water pollution induced water quality. Trade in virtual water form should be closely regarded into trade policies.

References

Feng K, Davis SJ, Suna L, Xin L, Guan D, Liu W, Liu Z, Hubacek K (2013) Outsourcing CO2 within China. Proc Natl Acad Sci 110(28):11654–11659

Feng K, Hubacek K, Pfister S, Yu Y, Sun L (2014) Virtual scarce water in China. Environ Sci Technol 48:7704–7713

Lin J, Pan D, Davis SJ, Zhang Q, He K, Wang C, Streets DG, Wuebbles DJ, Guan D (2014) China's international trade and air pollution in the United States. Proc Natl Acad Sci 111(5):1736–1741

Mekonnen MM, Hoekstra AY (2010) The green, blue and grey water footprint of crops and derived crop products, Value of Water Research Report Series No. 47, UNESCO-IHE, Delft, the Netherlands

Peters GP, Minx JC, Weber CL, Edenhofer O (2011) Growth in emission transfers via international trade from 1990 to 2008. Proc Natl Acad Sci USA 108:8903–8908

Zeng Z, Liu J, Savenije HHG (2013) A simple approach to assess water scarcity integrating water quantity and quality. Ecol Indic 34:441–449

Zhao X, Liu J, Yang H, Duarte R, Tillotson MR, Hubacek AK (2016) Burden shifting of water quantity and quality stress from megacity Shanghai. Water Resour Res 52(9):6916–6927

Printed in the United States
By Bookmasters